CARL DJERASSI

Born in Vienna but educated in the US, Carl Djerassi is a writer and professor of chemistry at Stanford University. Author of over 1200 scientific publications and seven monographs, he is one of the few American scientists to have been awarded both the National Medal of Science (in 1973, for the first synthesis of a steroid oral contraceptive— "the Pill") and the National Medal of Technology (in 1991, for promoting new approaches to insect control). A member of the US National Academy of Sciences and the American Academy of Arts and Sciences as well as many foreign academies, Djerassi has received 18 honorary doctorates together with numerous other honors, such as the first Wolf Prize in Chemistry, the first Award for the Industrial Application of Science from the National Academy of Sciences, and the American Chemical Society's highest award, the Priestley Medal. He is also an honorary member of the Royal Society of Chemistry.

For the past decade, he has turned to fiction writing, mostly in the genre of "science-in-fiction," whereby he illustrates, in the guise of realistic fiction, the human side of scientists and the personal conflicts faced by scientists in their quest for scientific knowledge, personal recognition, and financial rewards. In addition to novels (*Cantor's Dilemma; The Bourbaki Gambit; Marx, deceased; Menachem's Seed; NO*), short stories (*The Futurist and Other Stories*), and autobiography (*The Pill, Pygmy Chimps, and Degas' Horse*)—most of them written in London—he has recently embarked on a trilogy of plays with an emphasis on contemporary cutting-edge research in the biomedical sciences, which he describes in his web site as "science-in-theatre." "*AN IMMACULATE MISCONCEPTION*," first performed in abbreviated form at the 1998 Edinburgh Fringe Festival and subsequently (1999) in London (New End Theatre), San Francisco (Eureka Theatre) and Vienna (under the title *UNBEFLECKT* at the Jugendstiltheater), was broadcast by BBC Radio on its World Service in May 2000 as "Play of the Week." His second play, "*OXYGEN*" (co-authored with Roald Hoffmann) will open in 2001.

He is also the founder of the Djerassi Resident Artists Program near Woodside, California, which provides residencies and studio space for artists in the visual arts, literature, choreography and performing arts, and music. Over 1000 artists have passed through that program since its inception in 1982.

(There is a Web site about Carl Djerassi's writing at http://www.djerassi.com)

An Immaculate Misconception

Sex in an Age of Mechanical Reproduction

Carl Djerassi

www.djerassi.com

Imperial College Press

Published by

Imperial College Press
57 Shelton Street
Covent Garden
London WC2H 9HE

Distributed by

World Scientific Publishing Co. Pte. Ltd.
P O Box 128, Farrer Road, Singapore 912805
USA office: Suite 1B, 1060 Main Street, River Edge, NJ 07661
UK office: 57 Shelton Street, Covent Garden, London WC2H 9HE

British Library Cataloguing-in-Publication Data
A catalogue record for this book is available from the British Library.

This play is adapted in part from the novel "Menachem's Seed" by Carl Djerassi (Penguin-USA, New York, 1998)

AN IMMACULATE MISCONCEPTION

ISBN   1-86094-248-2 (pbk)

Printed in Singapore.

For Dale Djerassi
with love and thanks

Contents

Foreword

Bridging the Two Cultures

The gulf between the sciences and the other cultural worlds of the humanities and social sciences is increasingly widening, yet scientists themselves spend preciously little time in attempting to communicate with these other cultures. To a large extent this is due to the scientist's obsession with peer approval and the recognition that his tribe offers few incentives to communicating with a broader public that will do nothing for a scientist's professional reputation. Somewhat late in life, I decided to do something about illuminating the scientist's culture for a broader audience, and to do it through a tetralogy of novels in a genre I call "science-in-fiction"—not to be confused with science fiction. For me, a novel can only be anointed as "science-in-fiction" if all the science—i.e. <u>what</u> we do—and most of the idiosyncratic behavior of scientists—i.e. <u>how</u> we do it— described in it is plausible. None of these restrictions apply to science fiction. But if one actually wants to use fiction to smuggle scientific facts into the consciousness of a scientifically illiterate public—and I do think that such smuggling is intellectually and societally beneficial—then it is crucial that the facts behind that science be described accurately. Otherwise, how will the scientifically uninformed reader distinguish between what science is presented for entertainment and what is informative?

But of all literary forms, why use fiction—or drama? The majority of scientifically untrained persons are afraid of science. They raise a shield, the moment they learn that some scientific facts are about to be sprung on them. It is that portion of the public—the ascientific or even antiscientific person—that I want to touch. Instead of starting with the aggressive preamble, "let me tell you about my science," I prefer to start with the more innocent "let me tell you a story" and then incorporate realistic science and true-to-life scientists into the plot.

Scientists operate within a tribal culture whose rules, mores and idiosyncrasies are generally not communicated through specific lectures or books, but rather are acquired through a form of intellectual osmosis in a mentor-disciple relationship. Apprentice scientists acquire their "street smarts"—in some respects the soul and baggage of contemporary scientific behavior—by observing the mentor's self-interested concerns with publication practices and priorities, the order of the authors, the choice of the journal, the striving for academic fame—even culminating in Nobel lust. Each of these issues is loaded with ethical significance.

To me, it is important that the public does not look at scientists primarily as nerds, Frankensteins or Strangeloves. And because "science-in-fiction" or "science-in-theatre" deal not only with real science but more importantly with real scientists, I feel that a clansman can best describe a scientist's tribal culture and idiosyncratic behavior.

In our formal written discourse, we scientists never use the dialogic form—in fact we are not permitted to use it. Yet pedagogically, dialog is frequently much more accessible and—let us be frank—also more entertaining. The purest dialogic form of literature, of course, is drama. And if "science-in-fiction" is a rare genre, "science-in-theatre" is virtually unknown.

My interest in contributing to that genre was triggered by the success of Steven Poliakoff's **"Blinded by the Sun"** at the National Theatre, London (1996)—a play that received much publicity even in the scientific press. Illuminating in many respects very effectively some of the idiosyncratic aspects of a scientist's drive for name recognition as well as the competitive aspects of a collegial enterprise, it attempted to present in dramatic form the debate about "cold fusion" of a few years ago. Other sophisticated playwrights have used science for theatrical purposes where the science is incidental to the drama. Hugh Whitemore (**"Breaking the Code"**), Tom Stoppard (**"Arcadia"**), Friedrich Dürrenmatt (**"The Physicists"**), and Bertolt Brecht (**"Life of Galileo"**) are some earlier stellar examples.

In my projected trilogy, I am interested in the reverse approach: using the theatre for scientific enlightenment, *where the science is central rather than peripheral and impeccably correct.* As a model, consider Michael Frayn's **"Copenhagen"** (1998), a "science-in-theatre" play par excellence. Frayn makes no concession to scientific illiteracy. He calls upon quantum mechanics and the uncertainty principle for much of the scintillating interplay between two Nobelists, Werner Heisenberg and Niels Bohr.

Instead of selecting topics from contemporary chemistry or physics with their inherently complicated abstract terminology, I have turned to biology. More specifically, I chose recent, cutting-edge research in reproductive biology for four reasons: everyone can personally associate in one way or another with reproduction and sex; it encompasses an area of my professional competence; the terminology is relatively simple; and most importantly, such research is steeped with enormous ethical implications. To test these waters, I chose the ICSI technique as the scientific focus of **"An Immaculate Misconception,"** since in my opinion, ICSI—more than any other in vitro fertilization method—is contributing to the impending separation of sex ("in bed") and fertilization ("under the microscope").

I feel justified in assuming that most prospective readers of the present book or audiences intending to see a theatrical production of my play, will be unfamiliar with the term ICSI. Yet I am confident that—once having seen the injection of a single sperm into an egg in scenes 5 and 6 of **"An Immaculate Misconception"**—they will understand the ICSI technology and will never forget it or its ethical ramifications. If so, "science-in-theatre" will have bridged, however briefly, the widening gulf.

Program Note

Sex in an Age of Mechanical Reproduction

*"The technique of reproduction detaches the
reproduced object from the domain of tradition."*

(from Walter Benjamin, **The Work of Art in an Age
of Mechanical Reproduction**, 1936)

Impregnation of a woman's egg by a fertile man in normal intercourse requires tens of millions of sperm—as many as 100 million in one ejaculate. Successful fertilization with one single sperm is a total impossibility, considering that a man ejaculating even 1 – 3 million sperm is functionally infertile. But in 1992, Gianpiero Palermo, Hubert Joris, Paul Devroey, and André C. Van Steirteghem from the University of Brussels published their sensational paper in *Lancet*, **340**, 17 (1992), in which they announced the successful fertilization of a human egg with a <u>single</u> sperm by direct injection under the microscope, followed by reinsertion of the egg into the woman's uterus. ICSI—the accepted acronym for "intracytoplasmic sperm injection"— has now become the most powerful tool for the treatment of male infertility: over 10,000 ICSI babies have already been born since 1992.

This is the factual background of ICSI. But because "**An Immaculate Misconception**" is a play, all characters and events, though not the actual science*, are fictional—especially Dr. Melanie Laidlaw, ICSI's putative inventor. ICSI's ethical problems, however, remain even after the curtain has dropped.

---

*The film of an ICSI procedure shown in Scene 5 is based on an actual fertilization conducted by Dr. Roger A. Pedersen of the University of California, San Francisco, while that in Scene 6 was performed by Dr. Barry R. Behr of Stanford University.

Production History

The one-act version of "**An Immaculate Misconception**" (produced by William Archer) was first staged at the Edinburgh Festival Fringe from August 6 – 31, 1998.

Dr. MELANIE LAIDLAW	Jude Allen
MENACHEM DVIR	Michael Matus
Dr. FELIX FRANKENTHALER	Saul Reichlin
Director	William Archer
Lighting	David Babani
Stage Manager	Stephen M. Grasset
Original Music	Jessica and Alexander Boyd

The 2-act version, produced by David Babani, was staged at the New End Theatre, London, from March 17 – April 17, 1999.

The first American production was staged at the Eureka Theatre, San Francisco, from April 1 – May 2, 1999.

Dr. MELANIE LAIDLAW	Denise Balthrop Cassidy
MENACHEM DVIR	Peter Vilkin
Dr. FELIX FRANKENTHALER	Paul Sulzman
ADAM	Zach Kenney
FLORA MOTHERWELL*	Maxine Wyman

*(The role of Flora Motherwell was dropped in subsequent productions).

Director	Edward Hastings
Scenic Designer	Mikiko Uesugi
Lighting Designer	Jack Carpenter
Costume Designer	Rosey Bock
Stage Manager	Ed Fonseca

Original Music	Ron McFarland
Consulting Dramaturg	Bernard Weiner
Video Installation	April Minnich

The German version under the title **"Unbefleckt"** (translated by Bettina Arlt) premiered in Vienna on May 29, 1999 at the Jugendstiltheater am Steinhof.

Dr. MELANIE LAIDLAW	Susanna Kraus
MENACHEM DVIR	Alexander Strobele
Dr. FELIX FRANKENTHALER	Georg Schuchter
ADAM	Simon Schober

Director	Isabella Gregor
Scenic, Lighting and Costume Designer	Walter Vogelweider
Video Design	Heinz Wustinger
Consulting Dramaturg	Rosina Raffeiner
Video Installation	Actors-Service

BBC World Service broadcast a radio adaptation of **"An Immaculate Misconception"** as "Play of the Week" on May 7, 2000.

Dr. MELANIE LAIDLAW	Penny Downie
MENACHEM DVIR	Henry Goodman
Dr. FELIX FRANKENTHALER	Michael Cochrane
ADAM	Josh Brody

Director	Andy Jordan

GLOBALSTAGE (San Francisco, CA) produced and filmed a video version of **"An Immaculate Misconception"** (based largely on Carl Djerassi's adaptation for the BBC) for release in videocassette form effective June 2000.

Dr. MELANIE LAIDLAW	Eryn Maybruck
MENACHEM DVIR	Eric Wolfson
Dr. FELIX FRANKENTHALER	Mathew Dingess
ADAM	Edwin Day

Director	Lizbeth Pratt

Videocassettes (NTSC or PAL format) can be ordered from GLOBALSTAGE by fax (415-217-5819) or at www.globalstage.net.

Cast of Characters

Dr. MELANIE LAIDLAW: American reproductive biologist, late 30s, slender, athletic, with good looking legs (*relevant to scene 1*).

MENACHEM DVIR: Israeli nuclear engineer, 45 – 50, muscular and (preferably) hirsute. Speaks excellent English, but with very distinct Israeli accent.

Dr. FELIX FRANKENTHALER: American clinician and infertility specialist (late 30s to early 50s).

ADAM: Young teenager (17-year old in prologue, 13-year old in epilogue).

The action of the play takes place in an American location between 1997 and 1998.

PROLOGUE: Around 2014.

ACT 1, Scene 1: May 1997, sitting room of hotel suite on the occasion of a scientific Congress.

ACT 1, Scene 2: September 1997, Dr. Melanie Laidlaw's laboratory in an American Institute for Reproductive Biology and Infertility Research.

ACT 1, Scene 3: November 1997, dream scene in sperm bank.

ACT 1, Scene 4: January 1998, same location as *Scene 1*.

ACT 1, Scene 5: Sunday, February 8, 1998, same setting as *Scene 2*.

ACT 1, Scene 6: One hour later, same setting as preceding scene.

ACT 2, Scene 7: September 1998, same setting as preceding scene.

ACT 2, Scene 8: A few minutes later, coffee bar of Research Institute.

ACT 2, Scene 9: A few minutes later, same setting as scene 7.

ACT 2, Scene 10: One week later, same setting as preceding scene.

ACT 2, Scene 11: Early December 1998, Dr. Melanie Laidlaw's living room.

EPILOGUE: Thirteen years later (2011).

Technical Details

The two videos (see pp. 53–54 for still photographs) — provided by the author and to be shown on the rear screen or an appropriately sized TV-monitor in Scenes 5 and 6—depict an actual ICSI fertilization that needs to be coordinated with the dialog. (A sample sound dialog is included with one of the videos.)

The e-mail interludes can be projected in real time (*preferable*) or as intact texts following *Scenes 1, 2, 3, 4, 6 and 10*. Alternatively, only the e-mail headings may be projected with the text being heard through off-stage voices.

Prologue (Year 2014)

ADAM

(Pensive, touch of sadness)

The story you are about to watch happened about 17 years ago... before I was even born.

(Pause)

I can't remember exactly when I first heard my mother murmur in my ear, "Adam, my ICSI baby".... It sounded cuddly, the way she said "ICSI"... a new term of endearment. And then she'd always kiss me. *(Beat)*.

Another time I do remember hearing "ICSI," it came across very differently. Actually... I <u>over</u>heard it. She was talking about "<u>the</u>" ICSI baby on the telephone and it didn't sound cuddly. *(Beat)*. At least not to me. <u>The</u>, instead of <u>my</u>, sounded clinical... as if she'd converted me into an oddball or a milestone in medicine.

From that day on, I had a private 4-letter word. *(Beat)*. Not "Adam"... not "Life"... not "Love"... but *(beat)* "ICSI."

ACT 1

Scene 1

(May 1997, abroad, at a Scientific Congress): *MENACHEM DVIR and Dr. MELANIE LAIDLAW slouch post flagrante delicioso. Typical post-coital discourse: affectionate, tinge of guilt, touch of banter, curious yet private.*

MENACHEM
(*nuzzles her*)
> So you thought I <u>looked</u> married?

MELANIE
> You didn't look <u>single</u>. (*Beat*). You looked... (*searches for word*)... not loose enough. You aren't finger-branded, but I sensed some stamp of ownership.

MENACHEM
> So why didn't you ask... yesterday, at the opening session... or last night, in the sauna?

MELANIE
> I preferred not to know.

MENACHEM
> Because?

MELANIE
> Because if I had known—at that stage—that you were married... I mean known unequivocally... I wouldn't have... couldn't have.....

5

MENACHEM

I'm glad you did. By the way... you have great legs.

MELANIE

I know. But thanks anyway.

MENACHEM

And so smooth. When I saw your legs for the first time in the sauna last night, I knew I'd have to touch them. (*Beat*). You were the only one wrapped in a towel.

MELANIE

We Americans are primmer than Europeans. It's our puritan heritage... especially in saunas with strangers.

MENACHEM

Puritan... shmuritan! You may consider yourself too puritan to be sexy... but you're much too sexual to be prim. If I had to choose between sexy and sexual, I know what I'd pick.

MELANIE
(*quick*)

Who says you'd be the one doing the picking? (*Turns serious*). Do you believe that I've never done this before?

MENACHEM

Define "this."

MELANIE

Having ... carnal relations—

MENACHEM
(*grins*)

"Having carnal relations"—

(*Menachem attempts to continue speaking, but she leans over to put her hand over his mouth*)

MELANIE
—with a man I met only yesterday... some Israeli nuclear hotshot at this conference, who—

MENACHEM
(*Succeeds in pulling off her hand and laughs*)
... comes from the land of the Bible?

MELANIE
(*Irritable*)
You don't believe me? You think I make a habit of hopping into bed—

MENACHEM
(*Laughs*). How American! "Hopping into bed."

MELANIE
(*Irritable*)
How about... "I don't fuck men I don't know"?

MENACHEM
(*Gently*)
Melanie! Tsk, tsk.... Don't...
(*Tries to put index finger across her lips, but she bites it*)
Ouch!

MELANIE
All right.... so what would you say?

MENACHEM
Make love "with." Or maybe, "to."

MELANIE
And you prefer?

MENACHEM
"To."

MELANIE
Is that what we did?

MENACHEM
(*Very gentle*)
It was "with"... "To" is different. Someone has to take the initiative.

MELANIE
I see... and, of course, my virile Israeli wants to be the one—

MENACHEM
(*Plays with her hair or other gesture of affection*)
No, I don't... at least, not this time. (*Beat*) I think I'd leave it up to my puritan—

MELANIE
(*Quick, but softly*)
.... if there is a next time.

MENACHEM
There will be another time... there must be!

MELANIE
You're that sure?

MENACHEM
Yes... because you're not the type for one-night stands... as you Americans call them.

MELANIE

You really believe that? Honestly? Cross your heart?

MENACHEM

Crossing one's heart doesn't mean much to a Jew. But sure,
(*somewhat clumsily crosses his heart*)
I believe you—honestly.

MELANIE

How come?

MENACHEM

Guess.

MELANIE

(*Shakes head*)
No. You tell me... please. It's too early between us to even guess.

MENACHEM

All right. (*Beat*). I believe you, because it's also true for me.

MELANIE

You? You've never slept with a woman you barely knew?

MENACHEM

Well... (*beat*), not one I met only twenty-four hours ago.

MELANIE

Oh Menachem, I know so little about you.

MENACHEM

And I don't know much more about you... other than that you're a
scientist... or you wouldn't be here at this Congress.

MELANIE
You want to know what kind of science I do?

MENACHEM
No! It's not your science that interests me.... You can't make love to science. (*Beat*). Are you alone... I mean, in general?

MELANIE
Well... I have no husband and I have no children.

MENACHEM
So you want children?

MELANIE
Uh-huh.

MENACHEM
How old are you?

MELANIE
Guess.

MENACHEM
(*Reaches over to run his hand slowly over her face, like a blind person*)
Thirty-seven years plus or minus seven months.

MELANIE
(*Impressed*)
Not far off. So you see? I don't have too much time left... I mean for having children. But what about you?

MENACHEM
Do I want children? At one time, yes. (*Brusque*). But not anymore.

MELANIE
Am I getting too personal?

MENACHEM
Maybe. (*Beat*). Ask something else.

MELANIE
How old are <u>you</u>?

MENACHEM
(*Mock whisper*). Almost fifty. (*Louder*) Now it's my turn again.

MELANIE
Ask.

MENACHEM
What made you—

MELANIE
Hop into bed with you? Just because I have not made love with any man since my husband's death doesn't mean I'm not a sexual person.

MENACHEM
I'll vouch for that.

MELANIE
I don't want you to think I was a widow of opportunity. (*Beat*). But this scientist knows enough chemistry to recognize a unique reaction—one I've never experienced before.

MENACHEM
You're right about the spontaneous chemistry between us.

MELANIE
I said "unique."

MENACHEM
And the difference?

MELANIE
Spontaneous ones have a tendency to quickly fizzle out... unless you add something—

MENACHEM
Such as?

MELANIE
A chemist would say, you need more reagents... or maybe a catalyst.

MENACHEM
What kind?

MELANIE
It's too early to ask. Right now, the reaction is still sizzling... not fizzling.

MENACHEM
Maybe because I wanted it to sizzle, I didn't tell you—right then and there in the sauna—that I was married. But now you know everything.

MELANIE
Everything?
(*Long pause, looking anywhere but at MENACHEM*)
For a scientist, that's a meaningless word. You can never know everything. But you can learn enough to convince yourself to stop looking for more.
(*He starts to speak; she kisses his mouth shut and rolls onto him.*)

(*Languorously, erotically, slowly*)
This time... it will be "to."

END OF SCENE 1

E-Mail Interlude

After Scene 1 (following p. 12)

From: <mlaid@worldnet.att.com>
To: <mdvir@alpha.netvision.net.il>
Subject: Contact
Date: Sat, 24 May 1997 08:51:59

Dear M,

Since you gave me your e-mail address, can I assume nobody reads your messages?

From: <mdvir@alpha.netvision.net.il>
To:<mlaid@worldnet.att.com>
Subject: Privacy
Date: Sun, 25 May 1997 10:11:34

I'm the only one.

M.

From: <mlaid@worldnet.att.com>
To: <mdvir@alpha.netvision.net.il>
Subject: Explanation for tears
Date: Fri, 30 May 1997 21:01:45

My dear Menachem,

After our first night in Austria, my first night in bed with a virtual stranger, I thought: making love with a stranger is best, because there is no riddle and there is no test. But the last night, when you were no stranger anymore, showed that I was wrong.

You did write, but so little! Write more,

Melanie

Scene 2

(September 1997). Reproductive biology laboratory of Dr. Melanie Laidlaw at an American Infertility Research Center): *Spotlight on two stools and a lab table bearing typical biologist's lab paraphernalia: optional examples are petri dishes, pipette dispenser, rack of small tubes, perhaps a table-top centrifuge. The only indispensable item is a large microscope with double eyepiece, which is a key item in next scenes. Overall appearance somewhat untidy. FRANKENTHALER sits across from MELANIE on lab stool.*

MELANIE

Thanks for agreeing to join me. I'm now at a stage where I need scientific company. A clinical hot-shot like the eminent Dr. Felix Frankenthaler.

FRANKENTHALER
(*Bantering*)

What do you mean, someone "like" me? I thought you chose me because I was unique.

MELANIE

I didn't think flattery would hurt.

FRANKENTHALER

It never does.

MELANIE

But you <u>are</u> special: an infertility expert who treats patients... which is why I asked you. Yet you also keep up-to-date with basic science...

(*Mischievous tone*)… You're even known to dabble in the lab…. Anyway, we each bring something to the table that the other hasn't got. (*Beat*). Don't forget, your clinic is one of the best in the country.

FRANKENTHALER

You know it… I know it… but how many others? If your research pans out, that one publication will count for more than all the infertile patients I'll ever convert into parents.

MELANIE

I'm blushing.

FRANKENTHALER

I'm not flattering you. I'm just explaining why I accepted your offer so quickly.
(*Beat*). So how far are you?

MELANIE

A few more months and I'm ready to try fertilizing a human egg by direct injection with a single sperm!

FRANKENTHALER

Intracytoplasmic… sperm… injection (*Beat*)… ICSI. If it works, that acronym will be in the next edition of Webster's Dictionary! But just because you succeeded with hamsters, why not try it on monkeys next? What's the rush?

MELANIE

It's human reproduction and infertility, I'm interested in. (*Beat*). If your patients knew what I was up to in here… they'd be breaking down my door. Men with low sperm counts who can never become biological fathers in the usual way.

Mel realizes her endeavor is controversial.

FRANKENTHALER

My patients just want to fertilize an egg. They won't care if it's under a microscope or in bed... as long as it's their own sperm.

MELANIE

You're focusing on male infertility... that's your business. But do you realize what this will mean for women?

FRANKENTHALER

Of course! I treat male infertility to get women pregnant.

MELANIE

(*Amused*)

Felix, you haven't changed. You're a first-class doctor... (*pause*)

FRANKENTHALER

(*Bantering*)

But, but, but... Let's hear the but.

MELANIE

But... you look at everything through testosterone-tinted glasses.

FRANKENTHALER

(*Still affectionate banter*)

And what's my colleague's estrogen-etched view?

MELANIE

In the case of ICSI, that's easy—especially since my glasses aren't etched, but polished.

FRANKENTHALER

Aha!

MELANIE

Maybe that's why I see further than you. (*Beat*). ICSI could become an answer to overcoming the biological clock. And if that works, it will affect many more women than there are infertile men. (*Grins*). I'll even become famous. (*Beat*). We'll become famous!

FRANKENTHALER

(*Serious, even suspicious*)

What are you talking about? Sure... we'll be famous... world-famous... if that first ICSI fertilization is successful... and if a normal baby is born. But what's that got to do with (*slightly sarcastic*) "the biological clock?"

MELANIE

(*Leans forward, excited*)

Felix, in your IVF practice, it's not uncommon to freeze embryos for months and years before implanting them into a woman.

FRANKENTHALER

So?

MELANIE

So take frozen eggs.

FRANKENTHALER

(*Impatient*)

I know all about frozen eggs.... When you rethaw them, artificial insemination hardly ever works.... Do you want to hear the reasons for those failures?

MELANIE

Who cares? What I'm doing isn't ordinary artificial insemination... exposing the egg to lots of sperm and then letting them struggle on their own through the egg's natural barrier. (*Beat*). We inject right into the egg... (*Beat*). Now, if ICSI works in humans—

FRANKENTHALER
A big if.

MELANIE
(*Getting irritable*)
Felix... you're beginning to repeat yourself. It's not "if"... it's "when!"
And when is now! No hamsters... no monkeys... this is it! (*Motions to microscope*)

FRANKENTHALER
Melanie, slow down! (*Beat*). Failure is not going to get you famous.

MELANIE
I'm not talking about ICSI and fame... I'm talking about ICSI and motherhood.

FRANKENTHALER
Whose motherhood are you talking about?

MELANIE
(*Triumphant*)
Collective motherhood! Think of those women... right now, mostly professional ones...who postpone childbearing to their late thirties or even early forties. By then, the quality of their eggs... their own eggs... is not what it was when they were ten years younger.
    (*Becomes progressively more emphatic*)
So with ICSI, such women could draw on a bank account of their frozen young eggs and have a much better chance of having a normal pregnancy later on in life. I'm not talking about surrogate eggs—

FRANKENTHALER
Later in life? Past the menopause?

MELANIE

You convert men in their fifties into successful donors—

FRANKENTHALER

Then why not women? Are you serious?

MELANIE

I see no reason why women shouldn't have that option... at least under some circumstances.

FRANKENTHALER

Well—if that works... you won't just become famous... you'll be notorious.

MELANIE

I'll risk the notoriety. The fame, I'll share with you.

FRANKENTHALER
(*Mollified*)

Okay.... So we've got a new method of fertilization.

MELANIE

Think beyond that... to a wider vision of ICSI. I'm sure the day will come—maybe in another thirty years or even earlier—when sex and fertilization will be separate. Sex will be for love or lust—

FRANKENTHALER

And reproduction under the microscope?

MELANIE

And why not?

FRANKENTHALER

Reducing men to providers of a single sperm?

MELANIE

What's wrong with that… emphasizing quality rather than quantity? I'm not talking of test tube babies or genetic manipulation. And I'm certainly not promoting ovarian promiscuity, trying <u>different</u> men's sperm for each egg.

FRANKENTHALER

"Ovarian promiscuity!" That's a new one. (*Chuckles*). But then what?

MELANIE

Each embryo will be screened genetically <u>before</u> the best one is transferred back into the woman's uterus. All we'll be doing is improving the odds over Nature's roll of the dice. Before you know it, the 21st century will be called "The Century of Art." ∧ *Best will Survive*

FRANKENTHALER

Not science? Or technology?

MELANIE

The science of… A… R… T (*Beat*): assisted reproductive technologies. Young men and women will open reproductive bank accounts full of frozen sperm and eggs. And when they want a baby, they'll go to the bank to check out what they need.

FRANKENTHALER

And once they have such a bank account… get sterilized?

MELANIE

Exactly. If my prediction is on target, contraception will become superfluous.

FRANKENTHALER

(*Ironic*)

I see. And the pill will end up in a museum… (*Beat*)… of 20th century ART?

MELANIE

Of course it won't happen overnight.... But A... R... T is pushing us that way... and I'm not saying it's all for the good. It will first happen among the most affluent people... and certainly not all over the world. At the outset, I suspect it will be right here... in the States... and especially in California.

FRANKENTHALER

(*Shakes head*)

The Laidlaw Brave New World. (*Beat*). Before you know it, single women in that world may well be tempted to use ICSI to become the Amazons of the 21st century.

MELANIE

Forget about the Amazons! Instead, think of women who haven't found the right partner... or had been stuck with a lousy guy... or women who just want a child before it's too late...in other words, Felix, think of women like me.

END OF SCENE 2

E-Mail Interlude

After Scene 2 (following p. 22)

From: <mlaid@worldnet.att.com>
To: <mdvir@alpha.netvision.net.il>
Subject: Science news
Date: Tue, 16 Sep 1997 11:18:27

Dear Menachem,

Back then (how slow time passes!), you told me that one couldn't make love to science. I'm not so sure about that any more.

What I crave most now is working in the lab. My research is moving on express rails as if I was being rewarded for all the years of tedious progress and sliding backward. I have asked a first-class medical type to join me. So far, I have worked as the typical ivory tower scientist who wants to make her own reputation alone, but now I'm ready for teamwork.

This is just a quick note, because I've got to get back to the lab. I am starting to count the weeks toward our next conference in Austria!!!

The Puritan (who is getting less so every day)

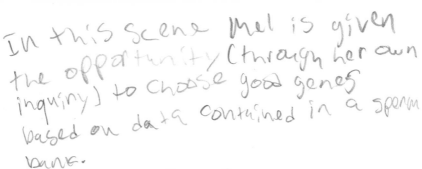

In this scene Mel is given the opportunity (through her own inquiry) to choose good genes based on data contained in a sperm bank.

Scene 3

(November 1997, American sperm bank office): *Spotlight on MELANIE. MALE VOICE is heard from offstage. Optionally, the entire scene can be staged as a dream.*

MELANIE
This is no ordinary bank.

MALE VOICE
You're no ordinary client, Dr. Laidlaw.

MELANIE
I want to check about withdrawals for research purposes.

MALE VOICE
You want to withdraw?

MELANIE
I'm just inquiring.

MALE VOICE
What size withdrawal?

MELANIE
Well... I need one spermatozoon.

MALE VOICE
Sorry... our minimal withdrawal is 80 million.

MELANIE
You see, I'm looking for a potential father.

MALE VOICE
Then a sperm bank is no place—

MELANIE
Sorry! I meant potential donor.

MALE VOICE
That we can provide.

MELANIE
I'd like to know about available choices.

MALE VOICE
Probably more than you can imagine.

MELANIE
Oh?

MALE VOICE
Try us. I'm sitting right in front of our computer... All I need to do is scroll.

MELANIE
How specific can I be?

MALE VOICE
Very. For instance... take hair:
    (*reads very rapidly*)

balding, thin, average, thick… and is it curly, wavy or straight? Or …
(*beat*)… dimples, cleft chin, Roman nose…. Is he right or left-handed?….
If there are freckles, are there few or many?
   (*Pause*)
I guess you're starting to get the picture. Or suppose you want to know
about the man's skin?
   (*Again speeds up*)
Very fair, fair, medium, olive or dark? And if the man checks "olive" or
"dark", he's got to check one of four boxes: light tan, dark tan, brown,
or black.

MELANIE
Enough! What about ethnic background?

MALE VOICE
Our current files list 80 choices.

MELANIE
That's too many for me.

MALE VOICE
You only need to pick one.

MELANIE
What about an Israeli?

MALE VOICE
Israeli? Let's have a look… I'm just scrolling on the screen.
   (*Pause*)
What do you know? Donor number 2062: Israeli… 5 foot 11, 165
pounds… straight black hair… majored in political science—

MELANIE
(*Laughs*)
Sounds promising.

MALE VOICE
> There are... let's see... 26 pages. Let me just give you the headings.
>   (*Starts speaking very rapidly*)
> Math skills, mechanical skills, athletic skills, favorite sport, hobbies, artistic abilities, favorite authors—

MELANIE
(*Astonished*)
> Authors?

MALE VOICE
(*Cool*)
> Why not? Suppose you noticed that your Israeli political scientist is reading Danielle Steele. Don't you think that tells you something about the man?

MELANIE
(*Laughs briefly*)
> I suppose so.
>   (*Quickly turns businesslike*)
> This is all very amusing... but what about genetic makeup?

MALE VOICE
> We check for Huntington's disease, Gaucher's disease, Wilson's disease, Crohn's disease... again, you get the idea. All donors are rigorously screened.... Additional tests may be requested... at your expense, of course. Full reports are available.

MELANIE
> Could I get a photo?

MALE VOICE
> This is a sperm bank, not a dating service. Our profiles will give you more data than most wives ever get.

MELANIE

I'm a widow... not a wife. And I don't want to adopt. I want to be a biological mother bearing my own child... and time is running out.

MALE VOICE

We'll provide a fertile donor... but no photos.

MELANIE

(*to herself*)

I was hoping I'd convince myself that an anonymous sperm donor would do. But no, I think I've got to hear the answers, not just read them. No... even that's not enough.... I guess I had to come here to find out I need to know the man.

MALE VOICE

A sperm bank may not be your first port of call then. (*Beat*) Romance is in short supply here. All we've got is billions of sperm, but no partners.

MELANIE

But I need one of each.

END OF SCENE 3

E-Mail Interlude

After Scene 3 (following p. 29)

From: <mlaid@worldnet.att.com>
To: <mdvir@alpha.netvision.net.il>
Subject: Desire
Date: Thu, 04 Dec 1997 11:32:28

Dearest Menachem,

Yes, I know. Another 29 days and I will have tasted you—in person. E-mail has its uses, but it also makes for impatience, and impatient I am!

I continue to be surprised by the intensity of my desire for you, and I am stunned by its persistence after such a long hiatus. A bridge connects, but it also separates—as does sexual pleasure. I have prided myself that what we had wasn't simply a one-night stand, but now I realize that a four-night stand is not much longer. Does the persistence of my desire—almost 8 months now—add to those days? Does it make this a 240-night stand? And can desire alone make something substantial out of what we did? Not mine alone, certainly. But if it were ours? OUR desire?

Your Puritan

Scene 4

(January 1998, living room area of a hotel suite at a scientific Congress).

The only light is a dim lamp on an end table or stray moonlight through a window. Upstage center is a partially open bedroom door through which off-stage sounds are heard of couple making love, culminating in the voice of a man coming to orgasm, sounding like an infant's birth cry, curiously unlike any other. Shortly thereafter, MELANIE is silhouetted in the doorway, barefoot, wearing nothing but a man's shirt, with a small object in her right hand. She moves quickly and quietly to the end table next to a sofa where a zippered toilet bag is seen in the dim light. It immediately becomes obvious that she is holding a used, extended condom. Carefully, in a time-consuming manner, she ties the open top of the condom into a knot and then reaches for the bag, trying (unsuccessfully) to unzip it. Finally, holding the condom between her teeth, she uses both hands to open the bag and removes a small, wide-mouth thermos bottle (or preferably, a Dewar flask), which she unscrews. She drops the condom directly from her mouth into the thermos, screws top on tightly with visible effort, then replaces it in the bag and closes the zipper. She is about to return to the bedroom when MENACHEM, sheet or blanket wrapped around him in toga fashion, appears in the doorway.

MELANIE
(*Recovers quickly from her surprise*)
    Greetings, Caesar!

MENACHEM
    What are you doing out here?

MELANIE
(*Coyly*)
You missed me?

MENACHEM
Of course I missed you.

MELANIE
So soon after—

MENACHEM
Especially so soon.

MELANIE
Well... here I am.

MENACHEM
But why did you leave?

MELANIE
(*Quickly interrupts*)
A woman's need... but you wouldn't understand. So don't ask.

(*Quickly leads him to sofa. They cuddle affectionately*)

MENACHEM
A penny for your thoughts.

MELANIE
Too cheap!

MENACHEM
Three pennies then.

MELANIE
Plus a kiss.

MENACHEM
It's a deal.
(Kisses her briefly)
And now the thought.

MELANIE
You call that a kiss? In exchange for one of my very own thoughts?
(Kisses him passionately, perhaps even climbing on him)

MENACHEM
(Finally disengages himself)
This better be some thought!

MELANIE
Remember when you asked me to join you in the sauna... with all
those other people... eight, long months ago?

MENACHEM
How could I forget? Isn't that where it all started?

MELANIE
(Turns pensive). And yet...

MENACHEM
Yet? That's a troubling word between lovers.

MELANIE
(Distances herself on the sofa)
    ... consider the sauna. There's a story about King Solomon and the
    Queen of Sheba—

MENACHEM
(*Dismissive*)
    That's old hat. It's in the Bible.

MELANIE
    I'm not talking about that tame one. There's one in the Koran. I bet you
    haven't heard it.

MENACHEM
    I'm a Jew.... Why would I read the Koran? But what does it say?

MELANIE
    According to it, Solomon received word that the Queen of Sheba was
    very beautiful, but he also heard that her legs and feet were hairy like
    the legs of an ass....

MENACHEM
    Is that right? What kind of an ass has feet for hooves?
        (*Reaches over to caress her naked feet, but she slaps his hand and
    moves away*)

MELANIE
    The ass in the Koran. (*Beat*). Solomon ordered the construction of a
    pond in front of his throne and had it covered with glass. When the
    Queen of Sheba approached his throne
        (*MELANIE acts out subsequent tale in pantomime*)
    and saw the water, she raised her long gown so as not to get it wet.
    Solomon couldn't help seeing that her legs and feet were handsome, but
    he disliked the hair upon her legs... The Koran says, "so the devils
    made for him a depilatory of quick-lime, wherewith she removed the
    hair."
        (*Reaches over to ruffle his hair*)
    Whereupon it came to pass that Solomon liked her ass.

MENACHEM
The Koran says that?

MELANIE
Well... not the bit about liking her ass. I made that part up.

MENACHEM
I like it... especially the bit you added. But what's the connection? Your legs aren't even hairy as I saw in the sauna...
   (*Reaches over to caress her legs to which she does not object*)
your towel wasn't long enough.

MELANIE
(*Triumphant*)
But don't you see? The sauna was your pond: you wanted to inspect me.

MENACHEM
So I've been found out! I plead guilty. (*Beat*). But since you tell me about Solomon, you should know what happened then.

MELANIE
I don't remember anything else.

MENACHEM
Do you know what it says in the Kebra Nagast?

MELANIE
The what?

MENACHEM
It's the Ethiopian Bible. According to it, the King hopped into the queen's bed or vice versa—

MELANIE
(*Playfully outraged*)

> What do you mean… "vice versa"? The Queen would never have done that!

MENACHEM

> Okay… so I made that one up. But the rest is true. Nine months later…
>> (*Embraces her and whispers*)
> The Queen bore fruit.

MELANIE

> Wasn't Solomon married when he met the Queen?

MENACHEM

> I don't remember. But what's that got to do with her becoming pregnant?

MELANIE

> Maybe she wanted a child and decided to just help herself to Solomon's seed—

MENACHEM

> Women! Well… at least that could never happen to me.

MELANIE

> Because you're smarter than Solomon?

MENACHEM

> Because I'm infertile… from a radiation accident. Severe oligospermia the specialist called it… always using Greek words when "you have too few sperm" would do perfectly well.

MELANIE
(*Taken aback*)

> So that's why you once said you can't have children anymore? I thought you'd meant your age. I didn't know you suffered from oligospermia.

MENACHEM

You still remember what I said eight months ago? (*Beat*) How time flies.

MELANIE

No... how slowly it passes. Eight long months. Even with e-mail, how many spontaneous chemical reactions can be kept going over such periods?

MENACHEM

(*A bit touchy*)

How could we help that? I'm in Israel and you're in the States.

MELANIE

And you're married... and I shouldn't allow myself the luxury... or is it the poverty?... of falling in love with a married man.

MENACHEM

Why not? I think I've fallen in love with you....

MELANIE

Men are different... they make love <u>to</u> women but basically, they're out to spread their seed.

MENACHEM

Not me.

MELANIE

What makes <u>you</u> different? Because you're Jewish?

MENACHEM

(*Slightly irritable*)

I already told you.... Because I'm infertile.

MELANIE

Infertility is relative... at least to this scientist.

MENACHEM

"Relative" infertility? Mine is pretty absolute.

MELANIE

Well of course, if you have <u>absolutely</u> no sperm, you're <u>absolutely</u> infertile. But that's pretty rare. Reproductive science is moving so rapidly nowadays...
(Catches herself before finishing sentence)....
But enough of science. As we both know, ours is not a scientific congress.

MENACHEM

So where does that leave us?

MELANIE

It leaves me waiting for the next scientific Congress... that my married lover always attends *(beat)* alone.

MENACHEM

Can you accept that?

MELANIE

I'm not sure. I've never had an affair before. I'm not the type... or at least I didn't think I was.

MENACHEM

But there is a bond.

MELANIE

There is one... or we wouldn't be here. But it's getting stretched.... Oh, Menachem, let's not risk breaking it.

MENACHEM
(*Gently*)
Give me something I can take back with me.

MELANIE
What?

MENACHEM
A strand of hair will do.

MELANIE
Help yourself.

MENACHEM
I will... later... after I've decided from where. (*Beat*). What would you like of me?

MELANIE
(*Moves away from him*)
Perhaps you've already given me all I've been missing.

END OF SCENE 4

E-Mail Interlude

After Scene 4 (following p. 39)

From: <mlaid@worldnet.att.com>
To: <mdvir@alpha.netvision.net.il>
Subject: Gift
Date: Sat, 07 Feb 1998 20:17:42

Dearest Menachem,

I am on pins and needles, because tomorrow is the great day in the lab —
perhaps the most important one of my life. Cross your fingers for me! If you bring
me luck, then you have given me the greatest gift you could offer.

In haste,

Your Melanie

Scene 5

(Sunday, February 8, 1998). *MELANIE, in surgical gown and cap, is sitting by the side of a small lab table with a standard ICSI setup consisting of microscope, micromanipulators and related gadgetry as well as a VCR unit connected to the microscope to project the image on a screen (or TV monitor). She sits perpendicular to the screen so that she can observe the images on the screen while pretending to look through the microscope.*

FRANKENTHALER
> Sorry I'm late.

MELANIE
(*Impatient*)
> Change into a gown. (*While he puts on gown, MELANIE continues adjusting the microscope*)

FRANKENTHALER
> Didn't anybody tell you that today is a Sunday, supposedly a day of rest.

MELANIE
> Felix. This is science, not religion.

FRANKENTHALER
> Oh yeah? If this works, don't think you won't be accused of playing God.

41

MELANIE

Let's worry about that later. Right now, I need steady hands.
(*Starts to put on plastic gloves*)

FRANKENTHALER

I'll say... with a pipette one tenth as thin as a human hair! But are your hands shaking? I'll go first if you like.
(*Playfully reaches for one of her gloves*)
Didn't I pass with flying colors practising on those hamster eggs?

MELANIE

(*Grabs glove back*)
Out of the question!

FRANKENTHALER

Melanie... I must ask again. Why don't you tell me whose eggs we're using?

MELANIE

I told you: it's my experiment.

FRANKENTHALER

But I'm your collaborator! Don't you think I'm entitled to know?

MELANIE

Of course... but all in good time.
(*Bends over microscope*)
We've got seven first-class eggs harvested—all from the same source. I'll do the first two. If all goes well, we'll break for lunch and then you can do the next two, as I promised.

FRANKENTHALER

Lunch? I think I'll avoid the chef's omelet.

MELANIE
No jokes please! Let's see how we do with this first one. Here we go.
   (*Puts on rubber gloves*)
Would you start the VCR?

FRANKENTHALER
Sure.
   Pushes the button and turns toward the screen. Both are completely silent as the screen lights up. MELANIE is hunched over the microscope, both hands manipulating the joysticks on each side of the microscope. She sits so as to be able to coordinate her words to action on the screen.
Ah... here we are. (*Startled*). God, this is low-grade stuff.
   (*For the initial image with lots of virtually immobile sperm, **have rapid ad-libbing** between MELANIE and FRANKENTHALER to match images on screen, **such as**)*

MELANIE
What do you expect from a functionally infertile man?

FRANKENTHALER
What? Are you crazy? Sperm from an <u>infertile</u> man? Why did you—?

MELANIE
   *When image of a couple of actively moving sperm appears, MELANIE, unwilling to disclose at this point source of sperm, interrupts. **Speed up tempo and excitement of ad-libbing** such as*
But these two <u>are</u> swimming—a good sign....

FRANKENTHALER
   (*Distracted from his concern as single active sperm appears at bottom of image, excitedly interrupts, though with sarcasm*)
Oh yeah, great... a real macho...
   (***Dialog has to be exactly coordinated with the events on the screen.***)

MELANIE

Okay... let's see how quickly I can catch this one. But first I've got to crush its tail so the sperm can't get away....

(*Gasps as sperm heads unexpectedly for capillary, then raises voice, shrilly, almost hysterically*)

Oh my God!... Look! Felix! Look!... It's heading straight for the capillary—head first!

FRANKENTHALER

Christ! He is <u>in</u> it! The wrong way! What now?

MELANIE

I'll have to kick it out and start all over.

(*Pause while she ejects sperm*)

Out you go! Bet you won't do that again when I crush your tail.

(*Quickly moves pipette toward sperm and sounds jubilant as the injection pipette crushes the sperm's tail*)

Gotcha!

FRANKENTHALER

Ouch!

MELANIE

Now comes the tricky part. I've got to aspirate it tail first.... As soon as I get close enough, just a little suction will do the trick.... Hah! Gotcha!

FRANKENTHALER

Not bad! Not bad at all.

(*Screen image displays the sperm, tail first, being sucked into the pipette. Image now shows MELANIE "playing" the sperm's head by moving it back and forward to demonstrate that she can manipulate it easily.*)

Quit playing with him! You've only got this single one!

MELANIE

I'm not playing with it. I just want to be sure that I can manipulate it at will. And why do you always call sperm "<u>him</u>"? Is it because the sex of a baby is always determined by the sperm?

(*Silence for a few seconds*)

Here we are.

(*Image of egg appears*)

Isn't <u>she</u> a beauty? Just look at her... here you are my precious baby... now stay still while I arrange you a bit... while I clasp you on my suction pipette... polar body on top....

(*Frankenthaler points to polar body*)

Like a little head. I want it in the 12 o'clock position.

(*Egg on screen is now immobilized in precisely the desired position for the penetration.*)

Felix, now cross your fingers.

(*He leans forward, clearly fascinated. Injection pipette containing sperm appears on image but pipette remains immobile.*)

FRANKENTHALER

(*Points to pipette on extreme right of image*)

What's the problem?

MELANIE

Nothing... it's just...

(*Pause, while image on screen shows injection pipette now aligned exactly in 3 o'clock position with respect to egg.*)

... doing it with <u>this</u> sperm into... <u>this</u>...

(*Does not finish the sentence as pipette penetrates the egg. MELANIE lets out audible gasp of relief.*)

FRANKENTHALER

(*Makes sudden start, as if he had been pricked*)

My God! You did it! Beautiful penetration!

(Image shows pipette resting within egg)
Now shoot him out!
(Points to sperm head in pipette)

MELANIE

Here we go.
(Image shows sperm head at the very end of the injection pipette, but it is not expelled. She aspirates it back and gives it a second push.)
Damn you! First you jump in when you aren't wanted and now you don't come out when you should! You've got to!
(At third attempt, one can clearly see the sperm head emerging on the screen from the pipette into the egg cytoplasm.)
Ah, that's a good boy.
(Carefully withdraws pipette without apparent damage to the egg.)

FRANKENTHALER

You did it! Look at him, just look at him! Sitting in there.
(Approaches image and points to sperm head on screen. Calmer voice.)
It's amazing. That egg looks... what shall I say?... inviolate, almost virginal.

MELANIE

(Looks up for first time from microscope)
It better not be... I violated it very consciously and tomorrow we'll see cell division.... Felix *(points to VCR)*, press the pause button, will you?
(He does and image of fertilized egg remains frozen on screen in full view.)

FRANKENTHALER

I guess I can uncross my fingers. But Melanie, whose crummy sperm are we using here? They were hardly moving.... You could hardly have chosen worse.

MELANIE
(*Bantering*)
> I could have picked sperm from a dead man.

FRANKENTHALER
> Dead man! Even if you're joking…(*shakes warning finger*), you'd be crossing ethical boundaries and not just biological ones.

MELANIE
> I'm not joking… I'm just speculating. If the sperm from a dead fertile man is aspirated within a few hours postmortem… maybe even after 24 hours… just so we still have some twitching sperm... one could preserve such semen for months, if not years and then still use it for ICSI.

FRANKENTHALER
> And you think that's okay? Using a dead man's sperm and, I suppose, a frozen egg of a deceased woman to generate instant orphans?

MELANIE
> No… I wouldn't go that far.

FRANKENTHALER
> But somebody else might.

MELANIE
> Kids need at least one parent.

FRANKENTHALE
(*Ironic*)
> I'm relieved to hear that. (*Beat*). So who is the father?

MELANIE
> There isn't any father in the usual sense of the word.

FRANKENTHALER
An immaculate conception?

MELANIE
You know, in a way that's true. There was no penetration of the woman, no sexual contact. In fact, at that moment, there was no woman, no vagina... nor a man (*beat*).... The only prick (*pause*)... was the gentle one by a tiny needle entering an egg in a dish, delivering a single sperm. (*Laughs*). Even that prick was provided by a woman. That process means nothing until the egg is transferred back into the woman.

FRANKENTHALER
Will you finally tell me who is this woman? I'll find out anyway in a couple of days.

MELANIE
Me.

FRANKENTHALER
Your own eggs? With all the mystery, I should have guessed.

MELANIE
Now, you know so. What you saw on this screen came from me. And so did the other six over there... (*points to petri dish*).

FRANKENTHALER
But why aren't you willing to wait until after we've established that ICSI works before using your own eggs? It's bad science... adding an emotional variable. It's crazy.

MELANIE
Why? Self-experimentation in medicine has an honorable tradition. Who was that guy who gave himself yellow fever?

FRANKENTHALER
It was Jesse Lazear... and he died from it.

MELANIE
I won't. I'll become a mother (*beat*)... and then famous.

FRANKENTHALER
Melanie... we're in this together. If you give birth, we publish together... and triumphantly. But if there's no baby... or even worse, a genetically damaged one... where does that leave things?

MELANIE
Felix, you have two children... and age doesn't matter to you. You can have more. But my time is running out quickly. Don't forget, I didn't freeze any of my young eggs.... Every year I wait now increases the risk.

FRANKENTHALER
(*Resigned*)
All right... all right. But did you really handpick this lousy stuff? <u>Why</u>, for heaven's sake. Why didn't you go to a sperm bank?

MELANIE
I've been to a sperm bank—

FRANKENTHALER
And?

MELANIE
Nothing happened. I just couldn't deal with an anonymous sperm donor. Period. I had to <u>know</u> the biological father of my child.

FRANKENTHALER
<u>You</u> cannot deal with an anonymous sperm donor? <u>You</u>—who with

ICSI—will have converted the average man—the donor of millions of sperm—into the provider of a single sperm? <u>You</u>—the mother of that intelligent New World as you called it not so long ago? And yet you have to <u>know</u> the donor of this bachelor sperm? What do you mean by "know"?

MELANIE

Overall health, even looks, may sound fine on paper... but the overriding factor would be intelligence... because I want my child to be the brightest kid around. But there are also intangibles you only sense about someone: kindness, wisdom, savoir faire, charisma... all kinds of personal things. In a sperm bank, you find sperm... not a man. With ICSI, I can consider everything.

FRANKENTHALER

I don't see how you can. Unless you are deliberately angling for an ordinarily <u>in</u>fertile man in your pool of potential fathers to prove the value of ICSI.

MELANIE

It's not a question of deliberately angling. It's just that you don't have to throw such a fish back into the water.

FRANKENTHALER

But that's crazy! Why take such a risk?

MELANIE

Because that pool of potential fathers... as you so aptly called it ... contained only one fish that interested me.

FRANKENTHALER

And you caught him?

MELANIE
And what if I did? Remember... these are <u>my</u> eggs we're injecting....

FRANKENTHALER
This first attempt at ICSI fertilization must be science... it can't be romance! Why is this man infertile? Have you looked into that? If there's some genetic information that we're missing, Laidlaw-Frankenthaler could give birth to who knows what.

MELANIE
I know what... I've done the work.

FRANKENTHALER
Oh, yeah. Did you get his consent?

MELANIE
Deep down, I know he'd like to be a father.

FRANKENTHALER
How deep?

MELANIE
(*Angry and loud*)
Stop it! This is not the time for such questions.

FRANKENTHALER
But a failure to disclose—

MELANIE
(*Completely loses her temper*)
Shut up! (*Beat*). Now my hands <u>are</u> shaking.
    (*Rips off gloves*)
Let's cool off and meet in half an hour.
    (*Exits*).

For a few seconds, FRANKENTHALER looks grimly into space, then turns around and turns on video image which again shows "dead" sperm seen earlier in scene. Suddenly reaches decision. Rushes to laboratory table, rummaging around (ostensibly looking for a condom). In desperation, he hastily grabs a fresh plastic glove and rushes off stage.

END OF SCENE 5

SPERM IMMOBILIZATION BEFORE INJECTION

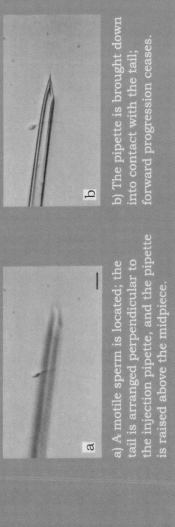

a) A motile sperm is located; the tail is arranged perpendicular to the injection pipette, and the pipette is raised above the midpiece.

b) The pipette is brought down into contact with the tail; forward progression ceases.

c) The pipette is drawn across the tail very quickly in the direction of the arrow.

d) The sperm is aspirated into the pipette tail first.

e) The sperm head is positioned close to the opening of the pipette.

INTRACYTOPLASMIC SPERM INJECTION (ICSI)

Holding pipette

Injection pipette pushing against zona pellucida

In each micrograph the sperm head is indicated by an arrow.

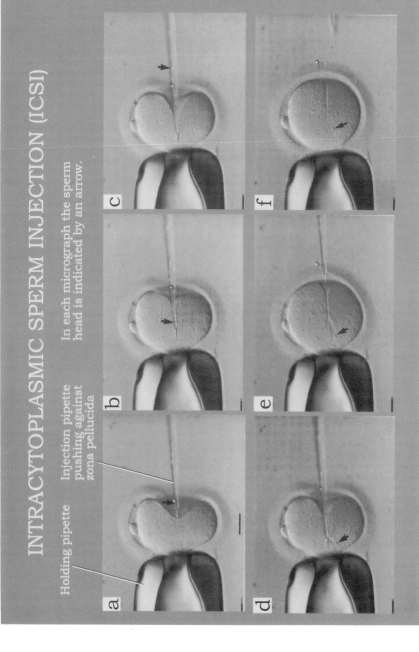

Scene 6

(Five minutes later). *FRANKENTHALER in surgical gown rushes into the lab, extended plastic glove (used in lieu of condom) in hand. Trying to open new syringe with one hand, he finally holds glove between his teeth while ripping off plastic cover of syringe. Quickly aspirates sperm from glove, drops a sample unto plate and places it under microscope. Turns on VCR. A new video image of very actively swimming sperm—quite different from "dead" sperm in preceding scene—appears on video monitor or screen.*

FRANKENTHALER
(*Looks up, gazes satisfiedly at image for several seconds, then murmurs loudly to himself.*)
    Now that's better!

        (*Puts on fresh plastic gloves and face mask and resumes microscope manipulation until the image on the screen shows quick aspiration of the sperm "tail first," avoiding MELANIE's initial "head-first" problem of scene 5. As image of new egg appears, he speaks loudly to himself.*)

Here you are, Melanie. Now let's see what we can do with you.
        (*As the second video sequence reaches point when injection capillary is about to penetrate egg, FRANKENTHALER again speaks to himself.*)
Relax, Melanie. I'll be very gentle... it won't hurt.
        (*Withdraws capillary slowly from egg*)
And now... bear fruit and multiply.

Suddenly, MELANIE enters.

MELANIE

Felix! What the hell are you doing?

FRANKENTHALER

(*Startled, jumps up, in the process knocking over lab stool. Quickly reaches for glove serving as condom substitute and rips off facemask. Stands up to block MELANIE's view of microscope.*)
I got back early... I thought I might as well ICSI the next egg. After all, we agreed—

MELANIE

... to proceed together.

FRANKENTHALER

But only one of us can manipulate—
(*Tries to throw glove into wastebasket, but misses*)

MELANIE

(*Brusquely interrupts him*)
Precisely! That's why I wanted to observe your manipulation!
(*Bends down to pick up glove from floor, but does not throw it into waste basket.*)

FRANKENTHALER

To be sure that I do it as well as you? I had no problems... as a matter of fact, my sperm sample didn't try to climb back into the capillary as your first one did.
(*Switches to conciliatory tone*)
Look... I'm really sorry I lost my temper earlier on. (*Beat*). But I guess some problems just take care of themselves. (*Beat*). Why don't we decide right now how many embryos to transfer back into you after we have injected the next two eggs?

MELANIE

(*Fidgeting with discarded glove in her hands*)
For the first time around, let's do two. The remaining ones we'll freeze.

FRANKENTHALER

As you wish. I'll just pick the two best looking ones.
(*Takes glove out of her hand and this time discarding it successfully in wastepaper basket.*)

MELANIE

No! I want to do the choosing.

FRANKENTHALER

Oh... but I'm an expert in choosing embryos. That's why I've got such a high batting average in my IVF practice.

MELANIE

I know. That's why I'm having you do the insertion. But not picking the embryos... that I want to do.

FRANKENTHALER

But why?

MELANIE

This is the first time in history that a woman has inserted a sperm directly into her own egg... completely on her own. Why have a man... even a man like you... pick the embryos for her? You might pick the two you injected. Don't you understand?

FRANKENTHALER

(*Disingenuously*)
But why should you care whose hand was on the injection pipette? It's still your own egg that's being put back into you. That's what should count... (*beat*)... that, and of course the sperm. Am I your partner or not?

MELANIE
(*Stubborn*)
> That's not the issue. What I did was the equivalent of having sex and
> fertilization all by myself.

FRANKENTHALER
(*Astonished*)
> And you find that attractive? Some people would call it gross.

MELANIE
> I didn't say it was attractive. It's just... a deeply emotional thing. A
> man wouldn't understand.

FRANKENTHALER
> It's time you stop playing multiple roles.

MELANIE
> What roles?

FRANKENTHALER
> That of a woman, who's obsessed with motherhood—something I see
> all the time in my practice—and simultaneously that of an ambitious
> scientist. (*Pause*).

MELANIE
> That's two roles.

FRANKENTHALER
> Two are enough to make my point.

MELANIE
> So get to the point.

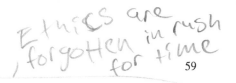
Ethics are forgotten in rush for time

FRANKENTHALER

As the ICSI scientist, you want to become famous.... That I can understand. But your hurry to become a mother is clouding your scientific judgment. That's where I come in... your ICSI partner... to look at the situation dispassionately. Selection of the best embryos for transfer should be done without emotion.... After the injection pipette is withdrawn from the egg and fertilization has started ... after ICSI has been completed... your role, as scientist, is over.

MELANIE

Oh, really? Who'll write the paper?

FRANKENTHALER

We'll both write the paper. But otherwise, you, as a lab scientist, must accept that as soon as the embryo enters your vagina, I, the physician am in charge.

MELANIE

So what are we arguing about?

FRANKENTHALER

Choosing the embryo is where the transition from you to me ought to start. Don't you trust me? Come on, Melanie! I'm your partner... by your own free choice.

MELANIE

I trust you.

FRANKENTHALER

In that case, I'll choose the two embryos I consider best before transferring them... with loving care... back into you.

(Pats her hand or other reassuring gesture)

MELANIE

Hand over the entire choice? No... I can't. It's not a question of trust...
it's a question of... (*Beat*)... what shall I call it? Emotional ownership.
But I'll split the difference with you. I'll pick the first embryo. (*Pause*).
And you can bet, it will come from one of the eggs I injected. And
since I'm the one who is labeling all the eggs and embryos, there won't
be any mistake! Then I'll let _you_ pick the second.

FRANKENTHALER

(*Reluctant*)

One embryo each?

MELANIE

(*Firm*)

A monumental concession on my part: one embryo each!

FRANKENTHALER

Okay.

(*Offers open palm, which she slaps; she then offers her palm, which
he slaps. Both then slap each other's palm, signifying agreement.*)

MELANIE

(*Approaches microscope*)

Great! Now that that's settled, let's finish ICSIing the other two eggs.

FRANKENTHALER

(*Quickly beats her to microscope*)

I might as well continue with the next one since I still have some sperm
here. You can then do the fourth.

(*Starts putting on plastic gloves*)

MELANIE

Good idea. I can check up on your technique.

(FRANKENTHALER *bends over the microscope*)

Let me start the VCR.

FRANKENTHALER
(*Quick reaction to avoid detection of his actively swimming sperm*)
No! Don't! Not yet!

MELANIE
(*Surprised*)
Why? I want to see how careful you are in injecting my egg.

FRANKENTHALER
That's okay. But first let me catch a good sperm. Then you can turn on the VCR and watch me do the rest.

MELANIE
(*Still surprised*)
Stage fright? Nonsense!

FRANKENTHALER
(*Disingenuously continues conversation while actually commencing work*)
It's a man's hang-up. It's brutal enough crushing the sperm's tail under the microscope without it being magnified and recorded.

MELANIE
(*Laughs*)
Poor squeamish Felix! You act as if it were your own sperm. (*Beat*). You behaved the same way when I caught the first one. You thought I was "playing" with <u>him.</u>
　　(*Good-natured*)
All right... I'll humor you. But the moment you've got that sperm in your capillary, I want to see the image on the monitor.

FRANKENTHALER
(*Relieved, while focusing on his work under the microscope*)
It's a deal. As soon as I get the little fellow into the capillary, you'll see it all.

(*Pause while he manipulates joysticks*)
Here she is! Now you can hit the VCR
   (*Brief view of egg from earlier video appears on screen.*)

MELANIE
(*Enchanted*)
   Another beauty!
      (*Long pause until image shows FRANKENTHALER completing the
      sperm injection.*)
   You did it!

FRANKENTHALER
   <u>We</u> did it.

END OF SCENE 6
END OF ACT 1

Act 2

E-Mail Interlude

Start of ACT 2 before Scene 7 (following p. 62)

From: <mdvir@alpha.netvision.net.il>
To:<mlaid@worldnet.att.com>
Subject: Once more
Date: Tue, 25 Aug 1998 09:42:04

Melanie— I HAVEN'T HEARD FROM YOU IN MONTHS. Didn't you get my last letter with my latest news?

I am leaving in a couple of days for the States on business. As soon as I get there, I'll stop at your lab.

Until then,

M.

From: <mlaid@worldnet.att.com>
To: <f.frank@compuserve.com>
Subject: Checking back
Date: Thu, 27 Aug 1998 51:34:19

Felix, my dear midwife,

Physically, I feel marvelous, though so huge that I wouldn't even mind a somewhat early delivery. But something else suddenly came up for which I need your assistance. Not as a clinical collaborator, but as a friend.

Give me a ring as soon as you're back in town.

Melanie

Scene 7

(September 1998, Dr. Melanie Laidlaw's laboratory; same setting as in Scene 5, except that teapot and cups are also visible). *MELANIE, seven months pregnant, sipping tea, sits by table. FRANKENTHALER enters, a transparent plastic bag of Chinese fortune cookies in his hand.*

FRANKENTHALER
(*Ebullient*)
    A very good afternoon to you, Dr. Laidlaw. Your midwife read your e-mail and is now reporting.
        (*Kisses her on cheek and waves bag with fortune cookies.*)
    What every pregnant woman needs in her seventh month: Chinese fortune cookies.

MELANIE
    Is this a friend's gift or a doctor's prescription?

FRANKENTHALER
    Why do you ask?

MELANIE
    I'm not crazy about fortune cookies.

FRANKENTHALER
    Doctor's prescription then. Go ahead... try one.
        (*He looks on as MELANIE breaks one open, reads message, puts it down and starts to open a second cookie.*)

You've got to eat one before opening another. Didn't the first one suit you?

MELANIE

What do you think?
  (*Shoves over paper slip*)

FRANKENTHALER

(*Starts reading proffered slip and laughs*)
I'll be damned! "Your problems are too complicated for fortune cookies."
But whose problems?

MELANIE

Whatever problems there are, they're ours.

FRANKENTHALER

Let's open another one.
  (*Quickly opens another, scans it, and then hands message to MELANIE.*)

MELANIE

"Fortune cookies are for the fools that buy them." (*Laughs*). What happened to "Confucius says?" Are you setting me up?

FRANKENTHALER

Serves me right for getting them at a Jewish delicatessen. Never again.

MELANIE

But why fortune cookies in the first place? Here I am, big as a blimp, 7 months pregnant... everything on schedule. We should celebrate with a cake, not this stuff. Laidlaw and Frankenthaler... three quarters toward the summit of ICSI. And no complications.

FRANKENTHALER

So far.

MELANIE
So far, so good. We aren't prophets... we're scientists.

FRANKENTHALER
Three-quarters up the mountain... or one quarter down. From now on... getting up to the top doesn't involve much science. Now all you need is a good doctor... like me... and a bit of luck.

MELANIE
Well, thank goodness. That explains the fortune cookies.

FRANKENTHALER
I admit being superstitious. (*Beat*). Like most scientists, including you: I noticed you used your favorite pipette for ICSI.
　　(*Shoves aside rest of unopened fortune cookies*)
But you're right, I should have brought a cake. A scientist shouldn't feed his superstition; he should stick to science.

MELANIE
So stick to it.

FRANKENTHALER
In that case, let's talk about the ICSI paper.

MELANIE
What's there still to talk about?

FRANKENTHALER
Pretty soon we ought to begin writing it.

MELANIE
I agree.

FRANKENTHALER
But there are still some preliminary housekeeping details....

MELANIE
Such as?

FRANKENTHALER
We agreed not to disclose the identity of the egg donor.

MELANIE
It's enough that you and I know.

FRANKENTHALER
But that doesn't apply to the sperm donor.

MELANIE
(*Laughs*)
You're still hung up on him?

FRANKENTHALER
Surely, as a physician... and as your partner seven months into this
project, I am <u>entitled</u> to know the source of that man's infertility? Is it
simply a low sperm count... and nothing else? In that case, the risk may
be tolerated for the very first ICSI procedure. But there may be many
other reasons for his manhood... if you permit a delicate phrase... to be
incomplete.

MELANIE
Incomplete manhood! Oh you sensitive men! (*Beat*). But what are you
driving at?

FRANKENTHALER
There are conditions where infertility is associated with serious genetic
disorders in the offspring. Cystic fibrosis, for one. The odds are high
there: one out of four.

MELANIE

I know all about that. You're talking about men suffering from congenital, bilateral absence of the vas deferens. (*Beat*). You see, I can also spout mumbo jumbo.

FRANKENTHALER

Let's not get technical right now. I'm talking about the principle of the thing.

MELANIE

And I'm talking about the technical principle: men with that condition have no sperm in their ejaculate. I can assure you that this man did ejaculate that sperm! I didn't have to aspirate it.

FRANKENTHALER

(*Angry*)

For God's sake, Melanie! Many other factors can lead to genetic abnormalities... and if such a baby is born, you can kiss the whole thing good-by.

MELANIE

Kiss what good-bye?

FRANKENTHALER

The ICSI paper. How could you justify sending it off if the result is some genetic...? (*Flustered pause*)... oddball... or whatever.

MELANIE

Felix! We're dealing with life... not a journal article!

FRANKENTHALER

Of course... and I apologize. I didn't mean it in such a callous way. After all, I'm also your obstetrician... or as you once called me: your midwife.

Life + science - where is the sanctity?

MELANIE

Then sound like one.

FRANKENTHALER

I promise. But now, as co-author of the paper, I've got to return to the question of the man's sperm. I presume he's on cloud 9, considering how few sperm of his I saw on the monitor. And now he's about to become the first ICSI father in history! What did he say when you first told him?

MELANIE

He knows nothing. I haven't seen him for months. He doesn't live here.

FRANKENTHALER

But...

MELANIE

He lives in Israel.

FRANKENTHALER

But...

MELANIE

He's married.

FRANKENTHALER

That, at least, I suspected... otherwise you surely would have told me about him. But you said you had his consent.

MELANIE

I said, he once wanted to be a father. You want consent for one miserable sperm?

FRANKENTHALER

Damn it, yes! I have a right to ask, at least in my capacity as a partner in a potential crime.

MELANIE

(*Exasperated*)

Crime!?

FRANKENTHALER

(*Irritable*)

Stop quibbling! How did you get a sperm sample without his knowledge?

MELANIE

What's the difference?

FRANKENTHALER

Come on, Melanie. How did you get it?

MELANIE

On one occasion... I kept the condom.

FRANKENTHALER

(*Openly sarcastic*)

And then you took it to the lab? Or did you have intercourse in the lab?

MELANIE

(*Angry*)

Don't be ridiculous. (*Pleading*). Felix, why go through all this?

FRANKENTHALER

Because deep down, I'm a very cautious physician who pays a hefty annual premium on his liability insurance. I've got to know the facts.

MELANIE

(*Resigned, yet impatient*)

All right. I had brought a small Dewar in my toilet bag and dropped the condom in it.

FRANKENTHALER

You stole it!

Ridding human life. What are the boundaries?

MELANIE

Felix! How can I steal something that the owner considers worthless? A used condom, for God's sake! (*Dismissive*). It was garbage... junk. Taking someone else's garbage or junk is not theft.

FRANKENTHALER

Junk and garbage are not the same. Junk is what you keep around. It only becomes garbage when you throw it away. In his body, a man's semen is mostly junk, not garbage. You, of all people, a reproductive biologist, should know that.

MELANIE

Stop acting so clever! If my egg is injected with an otherwise infertile sperm under a microscope and then put in a petri dish until cell division is confirmed, are you going to tell me that I'm now pregnant? Or that life has now begun? The egg has to be reintroduced into <u>me</u>, into <u>my</u> body... and it must implant in <u>my</u> uterus. Only then can we discuss the question of life. (*Beat*). Fertilization and pregnancy aren't synonymous.

FRANKENTHALER

As if I didn't know that! The whole abortion morality debate revolves around that issue.

MELANIE

abortion controversy

And for me right now, it's an issue of birth... not abortion. I'm speaking as a pregnant woman in her seventh month, who, let's not forget, invented ICSI. It's my method.

FRANKENTHALER
Our method.

MELANIE
I beg to differ! ICSI, as a procedure, was developed by me in animal models. And you know perfectly well how many years that took before I even got ICSI to work in hamsters! Converting this invention into human reality, producing a baby, that's our joint project.

FRANKENTHALER
Okay, okay....

MELANIE
So it's <u>my</u> method that transformed his garbage... sorry, I should've said junk... into something that could be used. That's not theft.

FRANKENTHALER
There are other ways of becoming a mother. You should have adopted. It would have been better all around... and ICSI would have remained science, clean and uncontaminated.

MELANIE
I'm not the adoptive kind... I'm possessive. I wanted my own biological child. (*Beat*). And then I met a man, who came out of the blue like a sexual angel... whom I could see as the biological father....

FRANKENTHALER
Why not beyond that?

MELANIE
Because he was married. (*Beat*). Because he considered himself infertile... and so did his wife. (*Beat*). So what should I have done?

FRANKENTHALER
People are known to leave one marriage for another.

MELANIE
I know. That's what he wrote recently: he's getting divorced.

FRANKENTHALER
Because of you?

MELANIE
Because his wife is pregnant.

FRANKENTHALER
(*Sarcastic*)
And how did your angel manage that?

MELANIE
You saw his sperm... (*Beat*). She got pregnant by another man.

FRANKENTHALER
So what else did he write? Or was it just about the weather in the Fertile Crescent?

MELANIE
Felix, don't be a wise guy. It doesn't become you. (*Beat*). He wrote he had business in the States and is coming over here.

FRANKENTHALER
To see you?

MELANIE
I guess so.

FRANKENTHALER
Guess? But haven't the two of you kept in touch since you saw each other last?

MELANIE
(*Guilty*)

We didn't. I know it bothered him, but what could I do? How could I keep up a correspondence with a former (*pause*)... lover without mentioning my pregnancy?

FRANKENTHALER

Now that he's getting divorced, what will you tell him?

MELANIE

Nothing.

FRANKENTHALER

But that's absurd! Would you have gotten yourself pregnant without him? Are you keeping him in the dark because you want the baby all to yourself?

MELANIE
(*Outraged*)

What? I want a baby of my own... my biological child... and you call that absurd?

FRANKENTHALER

Let me put it another way, because this is important for me.
    (*Speaks slowly with emphasis*)
Do you just want a child or do you want <u>his</u> child?

MELANIE

The latter, of course. Otherwise, why would I have selected his sperm sample? I admit that subliminally I may have been on the lookout for sperm. But every female does that... not just humans. With ICSI, I needed only one single sperm... but I had to know where it came from in the deepest sense.
    (*Suddenly looks startled*)

My God! I just realized that... here, I am... Melanie Laidlaw... who is about to achieve the creation of new life without sexual intercourse and yet I needed to acquire...

(*Change in tone and speed*)

that... single... precious... sperm through intercourse! The ultimate romantic scientist!

(*Pause*)

I have wanted a child more than anything else... even more than fame as a scientist.

(*Pleading tone*).

Felix, even if you disapprove... at least be fair.

FRANKENTHALER

I'm trying my best.

MELANIE

You promised to behave like a kind midwife, not a prosecutor.

FRANKENTHALER

(*Conciliatory*)

All right. So what do you want from me?

MELANIE

I can't possibly face him alone... I simply can't.

FRANKENTHALER

(*Gets up, tea cup in hand, slowly pacing up and down.*)

MELANIE

When he sees me like this (*points to stomach*) it has to be in neutral territory. (*Irritably*). Felix, stop wandering around. Sit down and help me.

FRANKENTHALER
(*Sits down*)
What sort of help do you want?

MELANIE
I don't want to hurt him when he comes this afternoon. (*Glances at watch*). He may be here any time. Just try to allude to my pregnancy in some subtle way.

FRANKENTHALER
It won't take him long to notice.
(*Points to her stomach*)

MELANIE
(*Beseeching*)
Please! It would make everything so much easier. Obviously, he would assume that there is another man in my life.

FRANKENTHALER
But why pick me? Because I'm the only one who knows? Have you never confided in anyone else? What about other women?

MELANIE
Women? I'm surrounded by people... but they're scientists... and in this place, they're mostly men. We talk about topics... we talk about problems... we talk about cases... and don't get me wrong... I do it because I like it. But we don't talk about ourselves. So will you help me... you, my midwife?

FRANKENTHALE
(*Moved*)
I'll give it a try.
(*Sound of sharp knock at the door. MELANIE looks startled. Freezes.*)

FRANKENTHALER
(*Whispers*)
> Well?

MELANIE goes slowly to the door just as there is a second knock and the door opens. MENACHEM, a bouquet of roses in his hand, and MELANIE nearly collide, surprising each other.

MENACHEM
(*Who has not yet seen FRANKENTHALER, loud and joyous*)
> Melanie!
> > (*About to kiss her when he notices her pregnancy.*)
> Oh!

MELANIE
(*Quickly*)
> Menachem, let me introduce to you to my friend, Felix Frankenthaler.
> > (*FRANKENTHALER approaches MENACHEM*)
> And this is Menachem Dvir, an acquaintance from Israel.
> > (*FRANKENTHALER extends his hand and MENACHEM in his confusion hands over the bouquet. A thorn pricks FRANKENTHALER's finger.*)

FRANKENTHALER
> Ouch!
> > (*He drops bouquet, which is retrieved by MENACHEM, who puts it on the lab table.*)

MELANIE
> Menachem, we're just having tea... and fortune cookies. Here, take a lab stool and join us.

MENACHEM
(*Picks up lab stool but doesn't take his eyes off MELANIE as he sits down.*)
> You look....

MELANIE
... very pregnant. (*Beat*). Here, have some tea... have a fortune cookie.

MENACHEM
Your face looks so different....

MELANIE
You mean it's also ballooned... like my belly?

MENACHEM
No, no... it's your expression. You look (*beat*)... I guess you look uneasy... and yet gratified. (*Beat*). A curious combination.

FRANKENTHALER
(*Tries to help MELANIE*)
Do you have fortune cookies in Israel?

MENACHEM
What did you say?

FRANKENTHALER
Fortune cookies... You break them open... They contain messages.

MENACHEM
(*Dismissive*)
Sure I know them. They're like horoscopes.
(*Addresses MELANIE*)
What else have you been doing since...?

MELANIE
It's been eight months. I guess a lot has happened to us both.

FRANKENTHALER
(*Addressing MENACHEM*)
> It won't hurt if you try one. Here, let me break one open for you.
>> (*Breaks open a fortune cookie and offers message to MENACHEM.*)

MENACHEM
> You people buy these things?

MELANIE
> I didn't... I don't believe in "messages."

MENACHEM
> So I gather. Neither do I.
>> (*Takes the proffered paper slip and throws it, unread, into the wastepaper basket some feet away, but misses.*)

FRANKENTHALER
> You need practice throwing trash.

MENACHEM
> What?
>> (*Again turns to MELANIE and studies her for several seconds.*)
> Your face has changed. Maybe I should've called it "blooming."

FRANKENTHALER
> "Germinating" is more appropriate. (*Awkward pause*). I think I'll leave you two alone. You must have lots to talk about. (*Exits*).

MENACHEM
> So you're pregnant.

MELANIE
> I'm afraid so.

MENACHEM
What's there to be afraid about?

MELANIE
Nothing... it's just a figure of speech. You know... one of our American expressions.

MENACHEM
Sure, like "hopping into bed." I've never forgotten that one.

MELANIE
(*Softly*)
Neither have I. (*Beat, then firmer*). But yes... I'm pregnant.

MENACHEM
I'm sorry... Sorry! I mean, I'm not sorry you're pregnant. I'm sorry I didn't know. When do you expect...?

MELANIE
In a couple of months.

MENACHEM
I wouldn't have barged in...

MELANIE
You didn't barge in. It's good to see you again, Menachem. You, at least, haven't changed. Not like me. So what brings you to the States? How long are you staying?

MENACHEM
Is he?

MELANIE
What?

MENACHEM
>   Is he the father?

MELANIE
>   Felix? (*Laughs*). No, of course not.

MENACHEM
>   You do look great... very pregnant... and, I suppose, very happy.... I think I better leave.

MELANIE
>   So soon? You've hardly said hello. I'm sorry you don't have more time.

MENACHEM
>   Time? That's not the issue.
>   (*Turns to face MELANIE*)
>   Good luck...with the baby. (*Beat*). And congratulate the father... whoever he is.
>   (*Exits*).

END OF SCENE 7

Scene 8

(A few minutes later): *Near coffee dispenser in hallway of Infertility Clinic. FRANKENTHALER and MENACHEM are in the process of sitting down, each carrying a cup of coffee.*

FRANKENTHALER
I thought I'd wait for you.
(*Looks him over before continuing*)
I noticed your surprise when you saw Melanie pregnant. I thought you were friends... (*beat*) and that you knew.

MENACHEM
Surprise is a bit of an understatement.
(*Pause, while he studies him*)
Do you know the father?

FRANKENTHALER
(*Stalling*)
Cream or sugar?

MENACHEM
Black. (*Beat*). So do you know him?

FRANKENTHALER
Sometimes you can never be sure who the real father is. But I've met a likely candidate.

MENACHEM
What does he look like?

FRANKENTHALER
(*Stalling*)
Oh, I'd say... (*Beat*) a bit like each of us.

MENACHEM
How old is he?

FRANKENTHALER
(*Pretends reflection*)
Oh, I'd say about your age... or... perhaps younger.
(*Studies him*)
How old are you?

MENACHEM
Pushing fifty.

FRANKENTHALER
Ancient. (*Chuckles*). Or let's say, maturing?

MENACHEM
Whatever. But what's he really like? Age and appearance in a man aren't that important to a woman.

FRANKENTHALER
I don't really know what attracted Melanie to that man.
(*Unable to restrain his curiosity*)
Where did you two first meet?

MENACHEM
(*Curtly*)
At a scientific congress.
(*Awkward pause*)

FRANKENTHALER
Melanie... she's an interesting woman, don't you think so?

MENACHEM
I'd have said, complicated.

FRANKENTHALER
That's all? Just... complicated?

MENACHEM
To me, complicated covers a lot of terrain. Melanie and I only saw each other for brief periods of time...
   (*Somewhat hesitantly*)
When we did, it was pretty intense... but....

FRANKENTHALER
Yes?

MENACHEM
Just when I thought I really understood her... when I thought I knew what made her tick... she... how shall I say it?... she escaped.

FRANKENTHALER
You're right! She does that sometimes... she just withdraws. A curtain drops.

MENACHEM
No... "withdraw" is not quite the same. "Escape" is sharper... you stand less of a chance of getting her back.

FRANKENTHALER
Are you good friends?

MENACHEM

Good? I don't know. You'll have to ask her. For me, good friends confide
in each other.

FRANKENTHALER

How true. (*Awkward pause*). Still, I had the impression that you two
were close.

MENACHEM

(*Surprised, somewhat suspicious*)
She talked to you about me?

FRANKENTHALER

(*Quick*)
No, not at all! It was the expression on your face... when you first
walked in...

MENACHEM

I see. But aren't you a friend of Melanie's? Actually (*laughs awkwardly*),
I thought you were more than that. (*Pause*). I thought you were the
father.

FRANKENTHALER

(*Curious and flattered*)
And why did you think that?

MENACHEM

(*Studies him carefully*)
You looked at ease... as if you had something in common. As if you
belonged.

FRANKENTHALER

(*Eager*)
We're partners... in reproduction. You might say we're the parents of a

procedure that will produce a baby within a couple of months. By the way, do you know what Melanie is working on... with me?

MENACHEM

We've never discussed her research. What's she doing?

FRANKENTHALER

It's very exciting, really. She's found a way of taking one single sperm and injecting it right into an egg. The procedure is called ICSI. It stands for (*pronounces slowly*) intracytoplasmic... sperm... injection.

MENACHEM

Big words.

FRANKENTHALER

For something very small: one little sperm.

MENACHEM

(*Intrigued*)

But isn't that just in the laboratory? Or does Melanie's ICSI work in the real world?

FRANKENTHALER

Oh, yes... at least it does so far. It resulted in fertilization... under the microscope. And I was the one who transferred the resulting embryo... actually two embryos... into the woman's uterus. Fortunately, one of them implanted... right at the first try. We'll have to see what happens. Technically, the first ICSI baby is still a fetus, but Melanie is already thinking about future applications... when sperm with Y and X-chromosomes can be separated. Once that is possible... and since you need only one sperm with ICSI... you can order a boy using a Y-sperm and a girl with an X-sperm. What do you think about that?

MENACHEM

Before talking about the future, let's stick to the present. You were talking about one baby. Then why did you transfer two embryos?

FRANKENTHALER

We do that in most IVF procedures for the sake of insurance... to be sure that at least one makes it. And while ICSI is a new form of fertilization... it's still in vitro fertilization.
     (*Studies MENACHEM*)
By the way... forgive me... but do you have children?

MENACHEM

What makes you ask that question?

FRANKENTHALER

A terrible habit... I'm the director of the fertility clinic here.

MENACHEM

Who doesn't want children... especially in Israel? But many years ago, I had a radiation accident... causing severe oligospermia. (*Bitter laugh*). That ended all questions of children.

FRANKENTHALER

Radiation?... You shouldn't be so categorical about that. I'm quite familiar with the effect of radiation exposure on sperm. It's very rapid, but depending on the radiation dose... and on the time elapsed... sperm counts can recover.

MENACHEM

You infertility doctors are professional optimists. That's exactly what the medics back home told me nearly 20 years ago. They were wrong.

FRANKENTHALER

With all the recent advances in treating male infertility... and especially Melanie's work...

MENACHEM

You mean it might work with me?

FRANKENTHALER

It might... especially if your wife is fertile.

MENACHEM

(*Sardonic*)

She's fertile all right.

FRANKENTHALER

(*Disingenuously*)

How can you be so certain? How would you know?

MENACHEM

(*Irritably*)

You're asking too many questions... even for a fertility doctor. Just take my word for it.

FRANKENTHALER

Sorry about that. I didn't mean to be intrusive.

MENACHEM

You were telling me about Melanie's ICSI.

FRANKENTHALER

If Melanie's procedure works... and we'll know that soon enough... it's tailor-made for men with low sperm counts... like you. (*Beat*). Of course, radiation damage can cause other complications.

MENACHEM

I know all about that. I'm a nuclear engineer.

FRANKENTHALER

In that case, I presume that even if ICSI were shown to work in a case like yours, you wouldn't try it.

MENACHEM

Why do you say that? I'd take the chance...

FRANKENTHALER

You would?

MENACHEM

You sound surprised.

FRANKENTHALER

I'm a very cautious man when it comes to genetic risks. I'd strongly advise against it. With sperm like yours...

MENACHEM

My sperm? Now where could you have seen that?

FRANKENTHALER

I said, sperm like yours. I know what sperm looks like after radiation exposure.

MENACHEM

What about Melanie? How worried is she about such risks? After all, ICSI is her baby.

FRANKENTHALER

Sometimes, women are bigger gamblers than men.

END OF SCENE 8

Scene 9

(A few minutes later. Dr. Melanie Laidlaw's laboratory, same as Scene 7.)
MELANIE stands pensively, holding MENACHEM's bouquet of roses in her
hand. Short quick knock. MENACHEM enters without waiting.

MELANIE
    Menachem!... I didn't expect you... again.

MENACHEM
    There's some unfinished business.

MELANIE
    Yes?

MENACHEM
(Points at her)
    How did you become pregnant?

MELANIE
(Flustered)
    What do you mean "how"? The usual process... a sperm penetrating
    my egg—

MENACHEM
    Don't you mean two sperm and two eggs? After all, weren't there two
    embryos? *(Beat)*. Was it ICSI?

MELANIE

How do you know about ICSI... and the eggs?

MENACHEM

I didn't until a few minutes ago. Your friend with the fortune cookie...
what's his name?... Frankenstein—

MELANIE

Frankenthaler.

MENACHEM

Whatever. He explained to me what one can do with ICSI. I could
hardly believe it. (*Beat*). So was it ICSI?

MELANIE

Yes.

MENACHEM

You experimented on yourself?

MELANIE

It's not unheard of in medicine... experimenting on yourself.

MENACHEM

Where did you get the sperm? A sperm bank?

MELANIE

No. I tried... but I couldn't do that.

MENACHEM

So... who's the father?

MELANIE

The father?

MENACHEM
(*Louder*)
    The father!

MELANIE
(*Stalling*)
    The father?

MENACHEM
(*Still louder*)
    Yes, the Father!

MELANIE
(*Completely loses composure*)
    The father?... the father?... the father?...
        (*Thrusts bouquet into his hands*)
    You! You're the father.

MENACHEM
    What?
        (*Pause for information to sink in. Then explodes*)
    And you tell me this now?
        (*Throws bouquet on the floor*)

MELANIE
    How could I've told you earlier?

MENACHEM
(*Yells*)
    I am the father and you have the nerve to ask,
        (*Attempts with falsetto to mimic her voice*)
    "How could I?"
        (*Almost screaming*)
    How could you <u>not</u>? In fact earlier... at the moment...

(*Suddenly stops. Approaches her as if he were to do her bodily harm, causing MELANIE to flinch*)
Wait a moment!... Wait! How did you get <u>my</u> sperm?

MELANIE
(*Tries to calm him while moving away from him*)
Menachem, wait. Please wait. One thing at a time. When I said, "How could I?" I meant knowing that you were married.

MENACHEM
I'm divorced.

MELANIE
Not then. You only just wrote that to me.... But I was sorry to hear of your divorce—

MENACHEM
(*Dismissive*)
Forget about my divorce! I want to hear how <u>you</u> got <u>my</u> sperm!

MELANIE
(*Reluctantly*)
I took the condom.

MENACHEM
(*Sarcastic*)
Ah yes... the condom! Your American obsession with safe sex!
    (*Curiosity taking over; lowers tone*)
But what did you do with it?

MELANIE
I put it in a Dewar flask containing liquid nitrogen.

MENACHEM
(*Sarcastic*)
You always travel with liquid nitrogen in your luggage?

MELANIE
Of course not.

MENACHEM
So that was planned?

MELANIE
It was... but not entirely. I didn't know about your infertility when I took the condom.

MENACHEM
(*Loud*)
Was everything planned? Right from the beginning? Were you just on a hunt for sperm?

MELANIE
How can you demean our relationship so much?

MENACHEM
You accuse <u>me</u> of demeaning? <u>You</u>, who reduced me to the ultimate dimension... one puny sperm... and then keeping it secret?

MELANIE
How can you say that? I acquired the sperm months after we first met.

MENACHEM
(*Screams*)
<u>Acquired</u>? Goddamn it... you <u>stole</u> it! And for what? An experiment? Or real fatherhood? And instead of giving me the greatest gift anyone could ever give me... you <u>hid</u> it?

(*Voice almost breaking with emotion*)
Why didn't you just <u>ask</u> if I wanted to be a father?

MELANIE
(*Voice almost breaking*)
Suppose you'd said "no?" I simply couldn't risk taking that chance... I wanted <u>your</u> child... not some anonymous man's sperm. So I preferred not to ask.

MENACHEM
No! You chose not to <u>tell</u>!

MELANIE
For God's sake! I was faced with the fact of your marriage—apparently a solid one.

MENACHEM
Some marriage!

MELANIE
I thought that was forbidden territory... that pursuing our spontaneous chemistry was as far as you could go... (*Beat*) ... or wished to.

MENACHEM
That marriage was well on the way to dissolution.... But as I said before, that's no concern of yours.

MELANIE
(*Quiet, yet firm*)
It <u>was</u> my concern... in more ways than one.

MENACHEM
Why?

MELANIE

When you wrote me about your divorce, I first thought your wife had discovered that we'd had an affair—

MENACHEM

(*Dismissive*)

That was months ago—

MELANIE

So what? It <u>was</u> an adulterous relationship.

MENACHEM

I told you then... and I tell you now: it was my problem—not yours. Ours was an affair between two consenting adults... evidently too brief to be converted into something more permanent. No one else was involved and nobody ever learned about it. Certainly not my wife... (*bitter*)... my former wife.

MELANIE

A man's rationalization.

MENACHEM

Most of us rationalize our actions. I repeat, nobody was harmed and my wife does not know about you to this day.

MELANIE

That's something to be thankful for. But you make it sound as if we just had a brief sexual affair and nothing more. Who's doing the demeaning now?

MENACHEM

I didn't say "brief." I said it was "<u>too</u> brief." That's a big difference. So what are you driving at?

MELANIE
(*Louder*)
> That brief sexual affair led to long-lasting consequences...
>> (*Turns away from MENACHEM*)
>
> Something that I've never regretted... though I've always felt guilty.
>> (*Turns back to him*)
>
> You still maintain I should've ignored that you were married? That I should've told you that you might become a father?

MENACHEM
(*Explosively*)
> Yes! Yes! Yes!

MELANIE
(*Defensive*)
> Why should I have raised your hopes? I had no way of knowing whether any of your sperm was viable. Furthermore, ICSI had never been tried on a human egg. I had no assurance that the embryo would implant in my uterus. And even now... 7 months after that event... I still don't know whether I'll give birth to a normal child....

MENACHEM
(*Sudden change in tone; concerned*)
> Stop saying that! Don't jinx it with all those "ifs" and "whethers."

MELANIE
(*Breaks into tears*)
> Thanks for saying that.

MENACHEM
(*Confused*)
> What do you mean?

MELANIE
> You're starting to sound like a real father.

MENACHEM
I <u>am</u> a real father!

MELANIE
I always knew that... but only by myself. Now we both know it. And that's one burden less... a very heavy one.

MENACHEM
(*Mollified*)
In that case, tell me.

MELANIE
Tell you what?

MENACHEM
About ICSI. What was it? What did you feel?

MELANIE
What do you mean?

MENACHEM
Well.... Women often claim that they know when they get pregnant... that the earth shakes... that the world moves... or something like that. But you must have actually seen the moment. (*Beat*). Was it like sex?

MELANIE
(*Gently*)
No... not like sex. Ours was magical, but that ICSI penetration... how can I explain?... It was unforgettable.

MENACHEM
And you injected two eggs?

MELANIE
Yes.

MENACHEM

And one is growing there?
(*Gently touches her belly*).

MELANIE

Yes... a boy of yours.

MENACHEM

(*Moved, caresses her belly*)
Oh, Melanie! (*Beat*). Remember Solomon and the Queen of Sheba?

MELANIE

What's that got to do with us?

MENACHEM

I never finished my part of the story. The Queen returned to Abyssinia
where she gave birth to a son, Menelek. Only years later, when Menelek
was a teenager, did she inform Solomon that he had a son. At least you
didn't wait that long to tell me.
(*Embrace or other gesture of affection between the two*)
So tell me.

MELANIE

Tell you what?

MENACHEM

Just tell me more... about ICSI.

MELANIE

Well... first, we injected your sperm into my eggs.

MENACHEM

(*Taken aback*)
We?

MELANIE

I did two... and Felix did two. And then we each picked one embryo and he transferred them back into me. One implanted... and now it's growing.

MENACHEM

What? Why did he choose one? (*Beat*). And why did you let him inject your eggs?

MELANIE

He's my partner.

MENACHEM

Okay. But did he do the second batch as well as you did the first?

MELANIE

I hope so.
(*Pats stomach*)

MENACHEM

Hope? Didn't you watch him do it?

MELANIE

Not the first one. I was out of the room.

MENACHEM

So whose sperm did he use?

MELANIE

Yours, of course.

MENACHEM

And you trust him?

MELANIE

I have no reason <u>not</u> to trust him.

MENACHEM

Absolutely none?

MELANIE

There are very few things in life that are absolute.

MENACHEM

In that case, ask Frankenthaler whose sperm he used.

MELANIE

(*Laughs*)

I can't ask him. I was only gone for half an hour or so. It sounds silly. Where would he have gotten it?

MENACHEM

(*Stubborn*)

I don't know.... (*Beat*). Maybe it is silly. Call it Jewish paranoia. Still... why don't you ask him? (*Beat*). Do me that favor.... Call it an early Father's Day present.

END OF SCENE 9

Scene 10

(One week later, Dr. Melanie Laidlaw's laboratory, same as scene 9.)
FRANKENTHALER enters, carrying a cake with a single unlit candle.
(***OPTIONALLY, this scene may be skipped.***)

FRANKENTHALER
(*Singing*)
> Happy birthday to you, happy birthday to you, happy birthday dear
> ICSI, happy birthday to you.

MELANIE
(*Interrupts laughingly*)
> Hold it, Felix! Whose birthday?

FRANKENTHALER
> This cake is for the ICSI paper. I went through your draft and my
> comments are so trivial, we might as well consider today its birthday.

MELANIE
> Well... why not? But first let me see your (*somewhat wary*) "trivial"
> comments.
>   (*Reaches for the manuscript, but he does not hand it over.*)

FRANKENTHALER
> But there's one non-trivial thing we ought to settle. Especially on the
> birthday.

MELANIE
Meaning?

FRANKENTHALER
Authorship.

MELANIE
What's there to settle? It's just the two of us. There are no other authors.

FRANKENTHALER
But whose name comes first? We've never discussed that delicate question.

MELANIE
You think it was delicacy that kept me from raising that issue?

FRANKENTHALER
So what was it?

MELANIE
To me, it was so obvious I didn't think there was anything to discuss.

FRANKENTHALER
Why not do it alphabetically?

MELANIE
Out of the question!

FRANKENTHALER
We could flip a coin.

MELANIE
I come first, because I thought of the idea. And then I reduced it to practice. Furthermore, it's <u>my</u> egg.

FRANKENTHALER

You are not going to say that your name comes first because you're the egg donor. In that case what about the sperm donor?

MELANIE

(*Laughs*)

You mean, add Menachem's name between mine and yours?

FRANKENTHALER

Of course not.

MELANIE

Felix, my name comes first. I wrote the manuscript and not you. Is that understood?

FRANKENTHALER

(*Grudgingly*)

Okay, okay.... Just kidding.

MELANIE

I'm not sure I believe that, but at least I'm glad to hear you say it.

FRANKENTHALER

In that case, let's light the candle and celebrate.

MELANIE

Let's.

(*FRANKENTHALER reaches in his pocket for a matchbox and strikes a match. At that moment, MELANIE continues*)

By the way, I never told you, but Menachem now knows that he's the father.

FRANKENTHALER

(*Stares at her speechlessly until the match burns his finger. Stamps it out.*)

What?

MELANIE
You look surprised. I told him the day you met him.

FRANKENTHALER
And?

MELANIE
He was pleased... once the initial shock wore off. But there was one question he asked... admittedly a silly question... when I told him about the ICSI procedure.

FRANKENTHALER
I bet I can guess.

MELANIE
(*Surprised*)
You can?

FRANKENTHALER
(*Pause*). He wanted to know what it felt like... when you injected the sperm.

MELANIE
(*Astonished*)
How did you guess?

FRANKENTHALER
(*Somewhat dismissive*)
Most men would think of that.

MELANIE
Did you?

FRANKENTHALER
Fleetingly... but mostly I was thinking of something else.

MELANIE
Do you remember what that was?

FRANKENTHALER
Yes.

MELANIE
Would you tell me?

FRANKENTHALER
(*Shrugs*)
Sure, why not?

MELANIE
So what was it?

FRANKENTHALER
I was still thinking of that miserable sperm when it suddenly surfaced on the monitor. I kept wondering what was so special about that mysterious man.... I suppose I was jealous for you to have taken such a chance.

MELANIE
And?

FRANKENTHALER
And nothing... that was it.

MELANIE
I also have a question. Can you guess that one as well?

FRANKENTHALER
I'm through, guessing.

MELANIE

You can't guess or you don't want to guess? Which is it?

FRANKENTHALER

Both.

MELANIE

When Menachem first raised the question, it seemed so preposterous... some kind of macho hang-up. (*Beat*). But now, I'm almost afraid to ask.

FRANKENTHALER

So don't ask. Some questions are best left buried.

MELANIE

But I promised Menachem I would.
    (*Walks toward him*)
Why didn't you want me to see the sperm capture on the video? Whose sperm did you use?

FRANKENTHALER
(*Rises and heads for the door*)
Melanie... if you have to ask that question, then surely you must know the answer.
    (*Exits*).

MELANIE
(*Long pause until full impact of remark sinks in*)
You bastard!

END OF SCENE 10

E-Mail Interlude

Before Scene 11 (following p. 110)

From: <f.frank@compuserve.com>
To:<mlaid@worldnet.att.com>
Subject: Final request
Date: Thu, 3 Dec 1998 27:02:09

Melanie—I understand why you may be pissed off, but I MUST see you once more IN PERSON!

Felix

Scene 11

(Early December 1998. Dr. Melanie Laidlaw's living room). *MELANIE (now slender) and FRANKENTHALER face each other.*

FRANKENTHALER
(*Agitated*)
> I don't get it. Your baby is one month old and you still won't...

MELANIE
> Do I have to spell it out for you? I... do... not... want... to... see... you... again! (*Beat*). <u>Ever</u>! I have nothing to say to you!

FRANKENTHALER
> For heaven's sake! Think about it! I had no choice.

MELANIE
> What made you think you were entitled to one? We were dealing with my eggs, <u>my</u> body, <u>my</u> prospective child—

FRANKENTHALER
> I'm not talking about <u>a</u> child... I'm talking about potentially thousands of children. That's what our experiment was all about! (*Beat*). One minute before performing the very first injection... in history... you confronted me on the monitor with a sperm sample that only a naïve optimist or a love-sick girl would consider suitable for ICSI. My injection was just as much for your personal benefit as for our success. Why weren't you willing to accept it as an insurance policy?

113

MELANIE
(*Sardonic*)
And you were just assuming the role of insurance agent who didn't
even inform me that I might be paying the premium? You decided that
your omnipotent sperm was just the ticket for me! What monumental
presumption! And not the vaguest hint... until I'm thirty weeks pregnant!
And only then when I asked you point blank.

FRANKENTHALER
Did you confide in the sperm donor what you planned to do with his
miserable sperm? Did you confide in <u>me</u>—the clinical colleague you
invited—when you hid the source of the eggs and the sperm until it was
too late? Yet you blame me for not confiding in you when I saw that
our ICSI experiment might be going down the drain? I had a 30-minute
window of opportunity when I was alone! What choice did I have? Take
a taxi to the closest sperm bank... (*beat*)... on a Sunday?

MELANIE
Get out of here!

FRANKENTHALER
Not until you've heard me out. I didn't come to complain that you
replaced me, your supervising obstetrician, in the last month. Unwise as
that may have been, it was your right to do so. But I also have some
rights—

MELANIE
You have none... as far as I'm concerned.

FRANKENTHALER
It was crazy choosing the sperm of a radiation accident victim for your
ICSI. Not only were you dealing with impaired fertility, you're facing
genetic risks.

MELANIE

That's why I insisted on screening of the embryos before transfer... that's why I went through an unprecedented battery of genetic tests by the end of the first trimester.

FRANKENTHALER

Well... you didn't need them with me. But God only knows what mutations are associated with the sperm of a man with a serious radiation exposure. How would anyone have known whether those tests were enough?

MELANIE

With genetics, enough is never enough... at least not until the entire human genome is deciphered. And that radiation accident was 20 years ago. (*Beat*). But wait a moment! Are you shedding all this genetic dandruff to tell me that I should gratefully accept my Adam—the first ICSI baby in history—as yours?

FRANKENTHALER

I wouldn't quite put it that way.

MELANIE

So what way would you put it?

FRANKENTHALER

Very simply. (*Beat*). I came to talk about the paternity—

MELANIE

Stop right there! You aren't going to tell me that you're demanding—

FRANKENTHALER

I've not come to talk about parental rights... The boy is your child.

MELANIE
(*Sarcastic*)
    So what <u>are</u> you talking about?

FRANKENTHALER
    Paternity. I want to know who belongs to whom—

MELANIE
    You just admitted Adam belonged to me... and to Menachem.

FRANKENTHALER
    God damn it! Just... let... me... finish! I'm talking about genetic information. Whose complement of genes does Adam have? Only then will you know what Menachem's role was... if any. (*Beat*). Every reasonable scientist would take that position.

MELANIE
    I would never have performed that ICSI experiment on my own eggs if it hadn't been for Menachem.

No Love only Science

FRANKENTHALER
    This is bigger than romance, Melanie. I'm looking at this as the co-author of the ICSI manuscript.

MELANIE
    You're worried about the names on a paper... whereas I'm talking about the parents of a living child.

FRANKENTHALER
    The two are closely connected. I came to request simple DNA analysis of the boy and the two putative fathers.

MELANIE
    Putative? (*Disgusted*)... Putative? The word alone makes me want to puke.

FRANKENTHALER

You can forget about that word the moment the DNA comparison is completed.

MELANIE

And I should approach Menachem with such a request? What makes you think you've even got the right to raise such a question?

FRANKENTHALER

Shouldn't he know that there were two eggs from one woman but sperm from two men? Now that a child is born? Don't you think he'd want to know whether Adam is his real son.... since only one of the embryos implanted?

MELANIE

Knowing about you is one thing. But asking him to participate in DNA analysis is something completely different. (*Beat*). Out of the question!

FRANKENTHALER

Okay... okay. Menachem isn't really needed. Comparison of tissue samples from Adam and me will be enough. If we don't match, Menachem is the father. And if we do—

MELANIE

You'll never find out, because I forbid it.

FRANKENTHALER

Why?

MELANIE

If I were sure that my son came from your ICSI...

FRANKENTHALER

Yes?

MELANIE

I couldn't cope with that prospect. I'm a parent before I'm a scientist.

FRANKENTHALER

I can respect that fear of knowledge... but only if you also acknowledge my desire for it. I've invested my brain and my heart... and my reputation in that ICSI experiment. (*Beat*). ICSI—the invention itself—must also mean something to you. You are its mother.

MELANIE

I know.

FRANKENTHALER

In that case, I must return to the paternity question—as the co-author of the ICSI paper.

MELANIE

Co-author? Hah! And what's that got to do with DNA testing?

FRANKENTHALER

Your son—the first ICSI baby in history—will be followed up... all through his life. You know that's a fact... Once the ICSI paper has appeared, the genie is out of the bottle. It can never again be recalled. (*Beat*). What if something turns out wrong with Adam? Suppose he turns out to be infertile? Is that a genetic problem or is it the ICSI procedure? How can we tell without DNA paternity testing? You seem to forget that before ICSI, men couldn't inherit infertility... it was uninheritable! But now?

MELANIE

You're turning my science against me. If Adam is infertile, he can use ICSI. Like father, like son.

FRANKENTHALER

But that's a prescription for a treatment, not an explanation for the cause. I repeat, how will you answer whether it was ICSI or the father's genes?

MELANIE

I won't, because I don't want to know the answer.

FRANKENTHALER

That's your privilege. But I must know the answer. You provide me with tissue specimens from Adam for the DNA typing and I'll take care of the test. The results will be my secret, but they'll settle once and for all the paternity of the first ICSI experiment. After all, that's why I did what I had to do. And that brings me back to the ICSI paper. I repeat... has it gone off to the journal?

MELANIE

It has. The day after Adam's delivery. By Federal Express. (*Sarcastic*). Do you want to see the receipt?

FRANKENTHALER

You did that without consulting me? That was totally unprofessional. What about my name?

MELANIE
(*Disdainfully*)

Your name is in the article.

FRANKENTHALER
(*Suspiciously*)

I don't like the way you said "in" the article.

MELANIE

I don't care whether you like it or not.

FRANKENTHALER
   Can I see it?

MELANIE
(*Disdainfully*)
   Sure.
      (*Rummages among papers on table or in briefcase, then shoves it in
      his direction.*)
   Here...

FRANKENTHALER
   (*Quickly grabs manuscript, looks at first page, stops and starts reading
   title aloud*)
   "First successful pregnancy after intracytoplasmic injection of single
   spermatozoon into an oocyte. A new treatment for male infertility. By
   Melanie Laidlaw, Ph.D." (*Beat*). Where the hell am I?

MELANIE
(*Simulated coolness*)
   Keep reading.
      (*Watches with evident satisfaction as FRANKENTHALER shuffles
      pages.*)
   It's on the last page.

FRANKENTHALER
   Where?

MELANIE
   Just before the bibliography.

FRANKENTHALER
(*Barely disguised fury*)
   I don't see a goddamn thing!

MELANIE
Read the acknowledgment.

FRANKENTHALER
(*Outraged*)
What!?

MELANIE
(*Reaches over to take last page out of his hand;*
reads in simulated precious tone)
"The author wishes to thank the National Institutes of Health for financial support, Dr. Gilbert Stork for advice on injection capillary design, Ms. Corazon de la Vega for assistance in oocyte retrieval, and (*beat*)... F. Frankenthaler for technical assistance."
(*Looks up at FRANKENTHALER*)
You see? You're in it... just as I told you.

FRANKENTHALER
How dare you?... Relegating me to a measly acknowledgment... and not even putting MD after my name!

MELANIE
You don't deserve it—not after what you have done—but I'll add "MD" after your name when I correct the proofs. Satisfied?

FRANKENTHALER
Do you really think I'll let you get away with this?

MELANIE
I don't know what choice you have.

FRANKENTHALER
I'll tell you what choice you have. Either you write a letter to the editor, admitting that a grievous error was committed and enclosing a new title page with both our names as co-authors... which I will mail—

MELANIE
    And if I refuse?

FRANKENTHALER
    I'll write to the editor informing him how you got the sperm sample.

MELANIE
    You want to blackmail me?

FRANKENTHALER
    Call it what you wish.

MELANIE
    I call it blackmail. When you injected your sperm into my egg without
    my knowledge or consent, you <u>raped</u> my egg. Do you want me to publish
    that? Rape and lack of informed consent?

FRANKENTHALER
    An ICSI injection into an egg isn't rape!

MELANIE
    The ICSI injection alone... without my knowledge... may be just a gross
    violation of informed consent—

FRANKENTHALER
    Consent? What about <u>your</u> misuse of Menachem's sperm?

MELANIE
    He has given consent.

FRANKENTHALER
    Because he thought he'd <u>become</u> a father? That would make it retroactive
    consent... a fine legal point in medical ethics. Or did he consent because
    he <u>thinks</u> he may be a father? But that doesn't <u>make</u> him one.

MELANIE

He is the other parent of Adam... I stand by him. Menachem considers himself the father. And the law recognizes him as such.

FRANKENTHALER

And who gave you that legal advice?

MELANIE

I don't need legal advice. We got married before Adam was born. The name on his birth certificate is Adam Dvir.

FRANKENTHALER

(*Taken aback*)

I'm surprised I didn't hear about your marriage.

(*Turns sarcastic*)

It's the first recorded failure of the lab grapevine.

MELANIE

It was private... We didn't think it was anybody's business... and least of all yours. (*Beat*). But back to my response to your blackmail. Your surreptitious ICSI injection may have been just a gross violation of medical ethics... if that particular embryo were discarded or didn't implant. But the birth of Adam converted it into potential rape... unless it is Menachem's embryo that implanted. (*Beat*). So you want a DNA test to prove that you are a rapist?

FRANKENTHALER

You're playing with words... or with insults. But where does that leave us?

MELANIE

Us? (*Louder*). Us? You stay in my footnote, but otherwise I won't snitch on you... if you don't snitch on me. You can call it a mutual deterrent.

MENACHEM
(*Offstage*)
> Melanie, where are you? The boy wants to be fed.   .

> (*MENACHEM enters, a teddy bear or other toy in his hand. He is surprised to find FRANKENTHALER. Turns to MELANIE.*)

> What's the fired midwife doing here?

MELANIE
> He showed up unannounced.

MENACHEM
> So what brought him here? And what are you two arguing about?

MELANIE
> Wait! Felix was about to leave. He was finished with what he came to tell me. Let's talk about it alone.

FRANKENTHALER
> I'm not leaving. Not until he hears my side of the story.

MENACHEM
> What's he talking about?

FRANKENTHALER
> Adam's paternity.

MELANIE
> Felix! I warned you!

MENACHEM
(*Waves away MELANIE*)
> That's okay... let him finish. What sort of paternity are you talking about?

FRANKENTHALER
The only paternity that counts.

MENACHEM
Adam has a father... and it's not you.

FRANKENTHALER
You're talking about the name on a birth certificate. I'm talking about
the pattern of a DNA gel.

MENACHEM
That's all there is to paternity? DNA patterns?

FRANKENTHALER
As far as ICSI is concerned... (*beat*), yes.

MENACHEM
(*Dismissive*)
You and your ICSI! (*Beat*). Paternity isn't just the provision of a single
sperm. It's also a human relationship... between father and son. (*Beat*).
Let me ask you a very simple question: suppose we performed DNA
analysis and you were shown to be the provider of that famous single
ICSI sperm.

FRANKENTHALER
Yes?

MENACHEM
Would you acknowledge Adam openly as your son? Would you support
him?
(*Quickly raises his hand*)
No! I withdraw that question. What I'm asking is... would you have
Adam <u>live</u> with you?
(*Throws teddy bear at him*)

MELANIE
(*Outraged*)
    Menachem! What are you asking?

MENACHEM
(*Waves her away, addresses FRANKENTHALER*)
    Well?

FRANKENTHALER
    No... I wouldn't go that far.
        (*Bends down to pick up teddy bear from the floor and places it
        carefully on the table or sofa. Turns serious and formal.*)
    Contrary to your insults, I'm no Frankenstein monster. I would not take
    a child from a mother who thinks of me as a rapist.

MELANIE
(*Heated*)
    You are one!

FRANKENTHALER
(*Quietly*)
    Am I?... Was I?... I thought only of <u>our</u> baby—but as an abstraction. I
    couldn't allow that first ICSI experiment—<u>our</u> ICSI experiment—to
    crash. You can call it what you wish—but rape it was not.

Exits.

MELANIE
    Good riddance.
        (*Turns to MENACHEM*)
    I'm sorry you walked into this. (*Beat*). What gall! Demanding DNA
    analysis... after what he did! I don't need DNA confirmation to tell me
    that Adam and you... together... belong to me. That we are the family!

MENACHEM
I know, but...

MELANIE
But? There's no but! There can't be any buts!

MENACHEM
But there is Adam.

MELANIE
Don't I know that? Who do you think I've been talking about?

MENACHEM
(*Calmly*)
Of course, about Adam, the baby...

MELANIE
And you.

MENACHEM
And me, yes.

MELANIE
So what are you driving at? That I'm not considering <u>him</u>...
    (*Points to door*)
that fraud?

MENACHEM
To him, Adam is ICSI... the product of a scientific invention. Of course, to you Adam is <u>our</u> precious baby... but he's also your ICSI baby. (*Beat*). Yet what about Adam when he's grown up... ready to lead his own, independent life? What will ICSI represent to him? A gift? Or a burden? What will paternity mean to him (<u>beat</u>)... <u>then</u>?

MELANIE

So what are you proposing?

MENACHEM

We'll take tissue samples... just his and mine... and have an independent lab perform the DNA analyses. Either we match or we don't. No one will see the results... no one but Adam.

MELANIE

And if they don't match?

MENACHEM

I'm Adam's father. Whether his father is also the sperm donor and how important that is to him are questions that will only occur to him when he learns the facts.

MELANIE

And when is that supposed to happen?

MENACHEM

That's for his parents to decide.

END OF SCENE 11

Epilogue (Year 2011)

ADAM

(Holds white envelope in one hand and a larger, brown manila envelope in the other)

We aren't really religious. My father still calls Mom "our Puritan" and he hardly ever goes to the temple. So I thought of my bar mitzvah more as some sort of coming-out party… than some serious religious initiation.

But my parents seemed nervous, which surprised me. They're not the nervous type… and most people at the service were joyous rather than serious. The situation sure turned serious though when we got home. That's when my mother gave me a sealed envelope.

(Lifts white envelope, which is torn open)

"I wrote it almost thirteen years ago," she said. "I would have kept it for another five or six before showing it to you. I think you are too young for this."

(Again lifts envelope)

Apparently, it was my father who convinced her otherwise. "It's the day of your bar mitzvah," he said, "when a Jew becomes a man. I think you're man enough to read what's in the letter." But then he handed me this.

(Lifts brown manila envelope in other hand)

He said it contained two samples… DNA samples… and the results of their comparison. Results that no one had seen as yet… although the tests were run at his insistence when I was still a baby.

I've never seen my father cry, but this time I saw tears in his eyes when he said, "Take your time," as he handed me this second envelope. "We'll wait for you upstairs." And then they left me with this... my bar mitzvah gift.

(*Points to both envelopes*)

I bet they're worried. I've been here for at least half an hour... but I can't go up yet. My mother was wrong: Thirteen is not too young to get this letter. (*Beat*). I've known for years that people call Mom, "the Mother of ICSI." I've been told that by now, there must be tens of thousands of ICSI babies all over the world—kids that would never have been born if it were not for her work. I guess no one will ever know that I'm *numero uno*, because it says here

(*Raises white envelope*)

that they'll never disclose my ICSI conception unless I announce it publicly. But what about that other man? Why did my father think that I should know about him?

"Some children in your situation—for instance, if their mothers went to a sperm bank—want to know who their biological father was," it says here.

(*Raises white envelope*)

I don't. (*Beat*). My mother didn't go to a sperm bank and my situation can't possibly be similar to anybody else's. Furthermore, she's convinced... "deep down within me," it says here... that my Dad is my father. So if I open the second envelope and the two samples don't match, does that mean my father stops being my father? I don't want to change my father and there's nothing I can do to change my genes anyway....

"Take your time," he said.

(*Long Pause*)

But I can't wait.

(*Pause, while he drops white envelope and starts—perhaps with hasty clumsiness because of Scotch Tape seal—to open the brown manila envelope*)

Not if I want to be a man.

(Takes out two longish X-ray strips, containing DNA patterns. First looks at one, then the other, then tries to line them up next to each other; finally superimposes one on top of the other and holds it up against the light—perhaps with back against the audience. As lights starts dimming, ADAM turns so that face is visible, showing a mixture of expressions: relief, shock, puzzlement.... then BLACKOUT).

END OF PLAY

Acknowledgments

An Immaculate Misconception was written in London. Its initial development was aided immeasurably by the generous advice of Martin Esslin, John Tydeman, and Alan Drury—all former members of the Theatre Department of BBC Radio—and by Nicholas Kent, Artistic Director of the Tricycle Theatre, who arranged the first staged, rehearsed reading on November 11, 1997 at his theatre under the direction of Erica Whyman with Michelle Fairley, Raad Rawi, Michael Cochrane, and Alexandra Lilley.

Further development in San Francisco was made possible through the Eureka Theatre (Bill Schwartz, Executive Producing Director, and Joe DeGuglielmo, Associate Artistic Director) who bridged two staged, rehearsed readings on March 24 and April 6, 1998 by a 10-day workshop with Cynthia Bassam, Charles Shaw Robinson, Peter Donat, and Carol Shoup-Sanders.

The ICSI procedure shown in Scene 5 is based on an actual fertilization conducted by Dr. Roger A. Pedersen of the University of California, San Francisco, while that in Scene 6 was performed by Dr. Barry R. Behr of Stanford University. Dr. Behr as well as the MRC Reproductive Biology Unit in Edinburgh and Prof. Salvador Moncada of University College, London generously provided microscopes and other laboratory equipment for the various theatrical productions in San Francisco, Edinburgh, and London.

Finally, it is my pleasure to acknowledge the generosity of N.V. Organon (Oss, Holland), who sponsored a special performance of **An Immaculate Misconception** on October 6, 1998 at the Yerba Buena Gardens Theatre in San Francisco on the occasion of the *World Congress on Fertility and*

Sterility, to Wyeth-Group (Münster, Germany) for sponsoring a special performance of the German version (**Unbefleckt**) on May 28, 1999 at the *Menopause Congress 1999* in Vienna, and the Bio-Gen-Tec Forum NRW 2000 for a similar sponsorship in Cologne on February 29, 2000.